Animal Husbandry and Nutrition

Animal Husbandry and Nutrition

Edited by:

Vijay D. Kele and Prafull Patil

Delve PUBLISHING

www.delvepublishing.com

Animal Husbandry and Nutrition

Vijay D. Kele and Prafull Patil

Delve Publishing
224 Shoreacres Road
Burlington, ON L7L 2H2
Canada
www.delvepublishing.com
Email: orders@arclereducation.com

ISBN: 978-1-77469-128-1 (Hardcover)

ABOUT THE EDITORS

Dr. Vijay Kele is presently serving as Associate Professor, Department of Dairy Technology and Food Technology, PIT, Parul University, Vadodara, Gujarat. He obtained his B. Tech (Dairy Technology) from MAFSU, Nagpur, M. Sc. (Dairy Science), MBA (Food & Agribusiness Management) & secured 04 gold medals from University of Mysore, Mysore and Doctorate in Dairy Extension Management from Nagpur. He having 11 years of Teaching and 03 years of industrial experience. He has published 04 books and 18 research paper in National and international journals of reputed, popular article 87, extension article 72 however 14 radio talk and 02 television live talk on Doordarshan. He has participated many National, State Seminar and symposiums. He has written 22 practical manuals for under graduate course for the benefit of students besides Lectures in farmer traning programme, organized 133 farmers scientist meets, organized 14 exhibitions. He has organized 73 webinars, 07 STTPs (one week), 12 Workshop and 01 FDP. As per research concern received grants for 04 projects from ICAR, Gujcost, SSIP (Govt. of Gujarat) However 02 Patents in his credits.

Dr.Prafullakumar V.Patil, presently working as a Farm Manager at Veterinary College, Udgir. He did his Master degree in Animal Nutrition in 2006 from MAFSU, Nagpur. He has qualified ICAR, NET examination in 2010. Now, pursuing PhD in Animal Nutrition at the College of Veterinary Science and Animal Husbandry, Anjora, Durg, Chhattisgarh. He is actively engaged in research, extension and clinical activities for the last 15 years. Also engaged in teaching from the last five years. He has published 37 research papers and 7 review papers in national and international journals along with 242 popular articles. He has received the best poster presentation award in the national symposium at PGIVAS, Akola (Maharashtra) in 2019. He has received a number of honours/awards for best extension work in the veterinary field. He has organized more than 21 trainings for field veterinarians; farmers etc. He has organized and participated in more than 150 farmers meet along with 45 animal health camps. He has authored two textbooks viz.'Handbook of Animal Nutrition' and 'Milk production management' in English and two books in Marathi. He has participated in more than 80 training programmes as a resource person and also conducted more than 130 demonstrations for field veterinarians, farmers and unemployed youths. He has also prepared an album of different feed ingredients and fodders. He has also acted as editor for 6 booklets on 'Profitable goat farming', 'Profitable dairy farming', 'Scientific poultry farming', 'Backyard poultry farming', 'Fodder production and management', Nutritional management of animals during scarcity period'.

TABLE OF CONTENTS

LIST OF FIGURES

LIST OF TABLES

LIST OF ABBREVIATIONS

AFRC	Agricultural and Food Research Council
AI	Artificial insemination
ARC	Agricultural Research Council
ATP	adenosine triphosphate
BC	Before Christ
BCE	before the Common Era
BP	before present
CAGR	compound annual growth rate
CH_4	Methane
CLR	common language runtime
CNS	Central nervous system
CO_2	carbon dioxide
CT	computerized tomography
DM	dry matter
DNA	deoxyribonucleic acid
DP	digestible protein
DSA	digital subtraction angiography
EID	electronic identification
ELISA	enzyme-linked immunosorbent assay
EPA	Environmental Protection Agency
ET	embryo transfer
EWS	early warning systems
FAO	Food and Agriculture Organization
GDP	Gross domestic product
GE	genetically engineered
GHG	greenhouse gas
GIS	geographic information system
GM	genetically modified

GNP	gold nanoparticle
GPS	global positioning system
GSDP	gross state domestic product
HPLC	High-performance liquid chromatography
IBERS	Institute of Biological, Environmental, and Rural Sciences
IETS	International Embryo Transfer Society
IITV	image intensifier TV
IoT	internet of things
IPCC	Intergovernmental Panel on Climate Change
IVF	in-vitro fertilization
LA	latex agglutination
LF	low frequency
MAI	marker-assisted introgression
MAS	marker-assisted selection
MOET	multiple ovulation embryo transfer technology
MRI	magnetic resonance imaging
NAIT	national animal identification and tracing
NE	net energy
NH_3	ammonia
NIR	near-infrared
NISP	number of identified specimens
NRC	The National Research Council
PCR	polymerase chain reaction
PPNA	pre-pottery Neolithic A
PPNB	pre-pottery Neolithic B
PUFF	pulmonary function
QEPAS	quartz unrivaled photoacoustic spectroscopy
RFID	radio frequency identification
RIA	radioimmunoassay
RS	remote sensing
SCC	self-contained controller
SSF	solid-state fermentation
TDLAS	tunable diode laser ingestion spectroscopy
TDN	total digestible nutrients

TI	thermal imaging
USDA	The United States Department of Agriculture
VFA	volatile fatty acids
VM	ventilator management
VOC	volatile organic compound
YBP	Year Before Present

GLOSSARY

A

Abridge – shorten (a piece of writing) without losing the sense.

Acetate – a salt or ester of acetic acid, containing the anion CH_3COO– or the group –$OOCCH_3$.

Anthropophily – a preference of a parasite or dermatophyte for humans over other animals.

Articulated – having two or more sections connected by a flexible joint.

Assimilation – the process of taking in and fully understanding information or ideas.

B

Biotechnology – is a broad area of biology, involving the use of living systems and organisms to develop or make products.

Bos-Indicus – is a scientific name of zebu cattle.

C

Carbohydrates – these are the sugars, starches, and fibres found in fruits, grains, vegetables, and milk products.

Colostrum – the first secretion from the mammary glands after giving birth, rich in antibodies.

Connotation – an idea or feeling which a word invokes for a person in addition to its literal or primary meaning.

Contagious – (of a disease) spread from one person or organism to another, typically by direct contact.

Contamination – the action or state of making or being made impure by polluting or poisoning.

Continental – forming or belonging to a continent.

Conventionally – in a way that is based on what is traditionally done or believed.

Conveyor – a person or thing that transports or communicates something.

D

Demonstrated – give a practical exhibition and explanation of (how a machine, skill, or craftworks or is performed).

Devised – plan or invent (a complex procedure, system, or mechanism) by careful thought.

Domestication – the process of taming an animal and keeping it as a pet or on a farm.

E

Endeavoured – try hard to do or achieve something.
Endemic – (of a disease or condition) regularly found among particular people or in a certain area.
Envisaging – contemplate or conceive of as a possibility or a desirable future event.
Epithelial – relating to or denoting the thin tissue forming the outer layer of a body's surface and lining the alimentary canal and other hollow structures.
Eutrophication – excessive richness of nutrients in a lake or other body of water, frequently due to run-off from the land, which causes a dense growth of plant life.
Excretion – (in living organisms and cells) the process of eliminating or expelling waste matter.
Externalization – giving external existence or form to.

F

Fabricated – invent (something) in order to deceive.
Faeces – waste matter remaining after a food has been digested, discharged from the bowels; excrement.
Feralization – wild, untamed, especially of domesticated animals have returned to the wild.
Foddering – the process of giving fodder to cattle or other livestock.
Forage – (of a person or animal) search widely for food or provisions.
Fragmentation – the process or state of breaking or being broken into fragments.

G

Garnered – gather or collect (something, especially information or approval).
Genome – the haploid set of chromosomes in a gamete or microorganism, or in each cell of a multicellular organism.
Genomic – relating to the haploid set of chromosomes in a gamete or microorganism, or the complete set of genes in a cell or organism.
Gestation – the process or period of developing inside the womb between conception and birth.
Grazing – a method of animal husbandry whereby domestic livestock is allowed to consume wild vegetations outdoor in order to convert grass and other forages into meat, milk, wool, and other animal products, often on land unsuitable for arable farming.

I

Imperative – of vital importance; crucial.
Incite – encourage or stir up (violent or unlawful behaviour).
Inclination – a person's natural tendency or urge to act or feel in a particular way; a disposition.
Incubated – (of a bird) sit on (eggs) in order to keep them warm and bring them to hatching.
Innately – as an inborn characteristic; naturally.

Insemination – the introduction of semen into a woman or a female animal by natural or artificial means.
Intravascular – situated or occurring within a vessel or vessels of an animal or plant, especially within a blood vessel or blood vascular system.
Isolation – the process or fact of isolating or being isolated.

L
Lactation – the secretion of milk by the mammary glands.
Lameness – (especially of an animal) unable to walk without difficulty as the result of an injury or illness affecting the leg or foot.
Legume – a leguminous plant (member of the pea family), especially one grown as a crop.
Livestock – domesticated animals raised in an agricultural setting to produce labour and commodities such as meat, eggs, milk, fur, leather, and wool.
Lucrative – producing a great deal of profit.

M
Mangles – destroy or severely damage by tearing or crushing.
Mastication – chewing of food.
Materialize – become actual fact; happen.
Meagre – (of something provided or available) lacking in quantity or quality.
Mechanization – the introduction of machines or automatic devices into a process, activity, or place.
Membrane – a thin sheet of tissue or layer of cells acting as a boundary, lining, or partition in an organism.
Monogastric – a monogastric refers to an organism that has a simple single-chambered stomach. Examples of monogastric herbivores are horses, rabbits, gerbils, and hamsters.
Morphometrics – the study of shape variation of organs and organisms and its covariation with other variables.
Multifaceted – having many sides.

N
Neolithic revolution – Also known as the first agricultural revolution. It is known as the wide-scale transition of many human cultures during the Neolithic period from a lifestyle of hunting and gathering to one of agriculture and settlement, making an increasingly large population possible.
Neolithization – The transition from an economy based on predation (hunting, gathering, fishing) to an economy based on the voluntary production of given plants and animals has been an essential process in the History of societies.
Neonatal – connected with babies who have just been born.
Nutrition – a field of study, the study of nutrients in food, how the body uses them, and the relationship between diet, health, and disease.

P

Paleodemographic – the description and explanation of biological adaptations, mortality, fertility, and migratory patterns of prehistoric populations, within the explanatory framework of evolutionary theory.
Paleolithic – refers to the prehistoric period when stone tools were made by humans.
Paleontological – the study of the history of life on Earth as based on fossils.
Parasitic – (of an organism) living as a parasite.
Pathogen – a bacterium, virus, or other microorganisms that can cause disease.
Phenotype – it refers to the observable physical properties of an organism; these include the organism's appearance, development, and behaviour.
Progeny – a descendant of the descendants of a person, animal, or plant; offspring.
Promptly – with little or no delay; immediately.
Prophylactic – intended to prevent disease.
Proximate – (especially of the cause of something) closest in a relationship; immediate.
Prudent – acting with or showing care and thought for the future.
Purebred – (of an animal) bred from parents of the same breed or variety.

Q

Quotation – a group of words taken from a text or speech and repeated by someone other than the original author or speaker.

R

Requisite – made necessary by particular circumstances or regulations.
Ruminant – and even-toed ungulate mammal that chews the cud regurgitated from its rumen. The ruminants comprise the cattle, sheep, antelopes, deer, giraffes, and their relatives.

S

Salable – fit or able to be sold.
Satiates – satisfy (a desire or an appetite) to the full.
Secretions – a process by which substances are produced and discharged from a cell, gland, or organ for a particular function in the organism or excretion.
Substrate – an underlying substance or layer.
Survival of the fittest – A mechanism of natural selection.

T

Therapeutic – relating to the healing of disease.
Transgenic – relating to or denoting an organism that contains genetic material into which DNA from an unrelated organism has been artificially introduced.
Tumors – a swelling of a part of the body, generally without inflammation, caused by an abnormal growth of tissue, whether benign or malignant.

U

Unfastened – open the fastening of; undo.

Unthrifty – not using money and other resources carefully; wasteful.
Urbane – (of a person) courteous and refined in manner.

V

Veterinarians – a person qualified to treat diseased or injured animals; a veterinary surgeon.
Veterinary – relating to the diseases, injuries, and treatment of farm and domestic animals.

W

Weaning – accustom (an infant or other young mammal) to food other than its mother's milk.

PREFACE

Animal husbandry, the branch of science, deals effectively with the practice of breeding, farming, and care of farm animals like cattle, dogs, horses, and sheep by humans for benefits. Alternatively, it is a branch of agriculture that involves livestock raising and selective breeding.

These days, farmers all around the world depend on animal husbandry for their livelihood. Animals, in addition to providing nutritious products to human, are also the major source of power for farmers as well as dryers. Thus, animal husbandry plays a key role in the rural economy.

This book attempts to bring the subject of animal husbandry and nutrition to the readers and help them realize the significance of this branch of agriculture. After reading the book, the readers will also equip themselves with the various kinds of uses and applications that animal husbandry are believed to serve and the possibilities that the researchers are pondering upon.

The book starts off by introducing the subject of animal husbandry to the readers and different aspects related to it, like history and objectives, so that the readers become comfortable with the term and develop a basic understanding before going forward into the book.

The 2nd chapter explains to the readers, the technology and science behind animal husbandry. It throws light upon various technological innovations such as the use of biotechnology along with the application of information technology in animal health.

Further, the 3rd chapter talks about the technologies that are used in animal husbandry in modern days. Also, the chapter narrates the method of breeding together with technologies improving productivity in the field of the livestock industry.

The 4th chapter throws light on the history and importance of nutrition in the field of animal husbandry, along with its basic introduction. Also, the chapter talks about archaeological approaches to early animal domestication.

The 5th chapter shifts the focus to the feeding standards for animals, their different types, and applications. Nutritional requirement for maintenance and growth is also described in detail. At the end of the chapter, some limitations of feeding standards are also given.

The 6th chapter takes the readers through the management of dairy farms, including record keeping, types of record-keeping, and its importance in dairy farms. Six key principles in dairy herd management will provide the reader with a basic understanding of dairy management.

The 7th chapter is related to the importance of cattle breeding along with their characteristics and management assessment. The last chapter throws light on the future aspects of animal nutrition and discusses the nutritional requirement for livestock.

I hope that the book interests the people from the subject area and that it can provide the readers with the best insights or discoveries that they may be looking for. We hope that this book can be a source of information and knowledge of animal husbandry and nutrition to the readers.

Chapter 1

Introduction to Animal Husbandry

CONTENTS

Animal husbandry, in general, refers to the raising of livestock and selective breeding, It is the care and management of animals in which the behaviour of animals along with their genetic qualities are developed to gain profit. Animal Husbandry is defined as the branch of science and art of management of common farm practices like scientific feeding, breeding, health care of common domestic animals aiming for maximum returns. All around the world, a huge number of farmers depend upon animal husbandry for their livelihood, so they require a lot of care as well as attention. In the beginning, the basic information about animal husbandry is provided along with its history and objective. Animal domestication is discussed towards the end.

1.1. INTRODUCTION

Animal husbandry ly a branch of agriculture that is concerned with the domestication of or cares for and breeding the animals like cattle, sheep, dog, pigs, horse, cattle, goat, and other similar creatures. Firstly, animal husbandry began in the Neolithic Revolution, which was around 10,000 years ago. There is the possibility that it might have begun much earlier.

It has been found that around 1.5 million years ago, human beings used f to to cook food for themselves. But according to the archaeological evidences that are being obtained is much far from the date of use of fire of cooking. This is about 12,500 years ago by the discovery of clay cooking started in East Asia and Mesopotamia.

According to the theory suggested by Charles Darwin, "Survival of the fittest," man is having mastered control over the animals. This means that man is havng a the capacity to design, develop, implement, and use the weapons that helps them to have control over the animals. Over the last 12,000 years, this theory has contributed to a larger extt in to the evolution the livestock.

Accordingthe historical evidnces, sheep and goats are considered to be the first species of animals that are being domesticatedthe humans for their use. This was further followed by domesticating of pigs. Near about 8,000 years ago, the cow was considered to be the last major food animal that was domesticateythe humans.

This has led to the introductionthe milk as a useful foodstuff. With this, there was also use of goat, sheep, camel, and reindeer milk. There was the use of other species like avian, fish, amphibians, and many arthropods for food.

As civilization developed, feed became more abundant, methods of caring livestock improved. Under man's selection they acquired qualities like rapid growth, better fat storage in body and increased milk production.

1.2. HISTORY OF ANIMAL HUSBANDRY

The Animal husbandry sector is considered to play a crucial role in ensuring the welfare of the rural population. It has been noticed thar the livelihood majority of farmers are dependent on Animal Husbandry. Supplementary employment and sustainable source of income are some of the other things that are being providy the livestock sector to many of the marginal and small farmers.

Therefore, it is one the most emerging and important sector which is leveraging the rural economy. This sector is additionally providing a continuous flow of essential food items like eggs, milk, and meat besides raw material like wool for the industrial areas, and manure.

Livestock rearing is also considered as an avocation with the increase in the productif the livestock products which is with the high export potential. The distributi of a livestock wealth is considered to be more egalitarian when compared to the land. Therefore, it is considered to be an important component in the poverty alleviation program from the livelihood and equality perspective.

Animal husbandry, as a component of the agricultural sector, is contributing about 3.92% to the Gross Domestic Product (GDP) (National Dairy Development board). With this, it also contributes about 5.5% to the employees of the labour forces.

Recently in few years, the livestock output has been growing at the rate of approximately 4 to 5% in a year. There is about 3.93% of the contribution of the livestock sector to the Gross State Domestic Product (GS and to the agriculture and allied activities; it is about 40.99% (Advanced Estimate at Current Prices).

There are different categories of wild animals. These animals are breeding better where they are found to be free than they do when they are captivated. These animals are found to have no common use for humans. For example, lion, rhinoceros, deer, tigers, etc., some animals are tamed like those who are caught from the wild and are being trained to be useful to humans.

Yak, gorilla, elephant, and chimpanzee are a few of the examples of Tamed animals asically caught from the wild and are later being trained to make them useful for human beings.

Few animals are being domesticated, which means that these kinds of animals are used at the homes and are considered to be easily bred and are completely looked at by humans. Some of the examples of domesticated animals are horse, cat, buffalo, dog, fowl, sheep, etc.

1.3. OBJECTIVE OF ANIMAL HUSBANDRY

Few objectives are important for animal husbandry and are as follows,

- Helps to increase the living standards of the farmers;
- Helps in develoal breed high yielder animal breeds;
- It helps in satisfthe need demand of the growing population for food;
- It helps in the proper management of the domestic animals;
- It leads to the high production of eggs;
- It leads to the high production of milk with improved quality;
- It gives in high yield of Fish production;
- It gives high meat production;
- It also helps in the utilization and systematic disposal of the animal waste;
- It helps in maintaining a healthy environment.

1.4. ANIMALS IN THE NEOLITHIC

It is believed that there is a tendency which is especially within the secondary literature, which is within accounts of the Neolithic revohis basically emphasizes the broad evolutioerns in order to treat Southwest Asia as a homogenous entity.

It also equates the Neomal economics economy which is with those who are dominated by the 'barnyard complex' of domesticated animals such as goat, pigs, cattle, and sheep (Figure 1.1) (Banning, 1998; Bellwood, 2005; Cauvin, 2000; Cunliffe, 2008, p. 88; Guilaine, 2007).

Figure 1.1. Neolithic revolution.

Source: Image by Lincom.org.

According to recent research that has basically bougemporal pattern and geographic pattern for the domestication of the below mentioned animals into a more detailed focus (Vigne et al., 2011; Zeder, 2008b, 2011).

It has therefore become increasingly clear that this narrative is not accurate. It is also considered that the transition from hunting to husbandry was neither homogeneous nor it is rapid across Southwest Asia (Asouti and Fuller, 2012; Fuller et al., 2012).

However, the Neolithic economies present in this region which is far from the representations of the homogenous evolutionary stages. It, therefore, reflects the panoply of the divergent responses towards the complex local scale requirement of the task scheduling, social needs, gender politics, and the cosmologies.

Effects of Neolithic Revolution

The Neolithic Revolution led to a huge number of people establishing permanent settlements supported by agriculture and farming. It surfaced the way for the innovations of the ensuing Bronze and Iron Age when developments in creating tools for farming, art, and wars swept the world and brought civilizations together through trade as well as conquest.

These are believed to feature themselves in order to be defined by the combination of local histories, which are found to be embedded within the limitations and opportunities often being provided by the local environments.

1.5. ZOOARCHAEOLOGY AND ANIMAL DOMESTICATION

Animal domestication is basically described as a long-term process. In this process, humans are applying intensive, intentional, and consistent control over the animal populations, which as a result brings a significant change in the breeding, diet, phenotype, and even in mobility (Bökönyi, 1969; Ducos, 1978; Meadow, 1989).

However, humans are found to interact with the animals in different ways, and therefore, there is no generally agreed upon the making of the boundary at the beginning of a domestic relationship or at the end of the other type of relationship which is associated with the hunting, taming, pet-keeping, wildlife management, and commensalism (Figure 1.2) (Russell, 2010; Vigne et al., 2003, 2011; Zeder, 2012b).

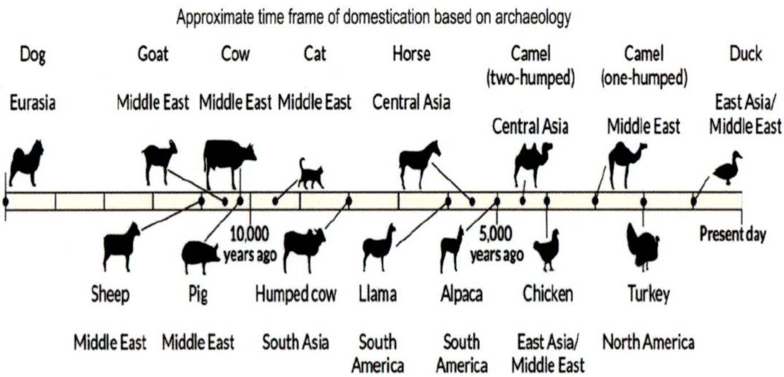

Figure 1.2. Animal domestication.

Source: *Image by WordPress.*

There are more intensive relationships that involve human control that is over movement, breeding, and feeding of the animal populations. This is found to be identified in the archaeological record with the help of slaughter profiles that are targeting young males, isotropic evidence for the foddering.

There are other forms of transport of animals and dietary manipulation which is outside their natural habitats (Bökönyi, 1969; Makarewicz and Tuross, 2012; Zeder and Hesse, 2000).

'Domestication syndrome' in animals is the result of intensive management by humans, which is largely a persistent and intentional manipulation of the animal population's reproduction, their isolation from the wild population, access to food, and their mobility which a while over the generations causes several phenotypic and behavioural changes in these animals (Figure 1.3) (Hammer, 1984).

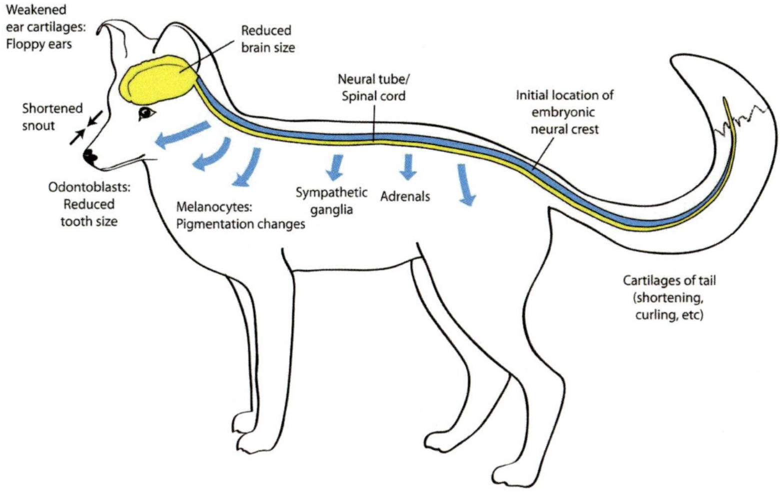

Figure 1.3. Domestication syndrome.

Source: Image by Genetics.

The primary focus of many archaeozoological studies concerning animal domestication has been to identify the features of this specific syndrome whereby pedomorphic traits like a reduced body and brain size, changes brought about to the morphology of the horn as evident amongst the ungulates as well as the snout's foreshortening have been studied (Hole et al., 1969).

At the same time, it is evident that only the later stages when the human control over the animal population intensifies are visible when the focus is on studying the appearance of 'morphological domesticates'(Vigne et al., 2012).

When the management is not very intensive and during the earlier phases of the process involved in domestication, human management, and the morphological changes do not always vary concerning the same variables due to which the approaches towards this study have lately been emphasizing upon the treatment of these two factors as independent analytical units (Hadjikoumis, 2012; Vigne et al., 2011; Zeder, 2006).

On the other hand, the variations in the morphological changes can be taken to be dependent on a wide range of issues like the genetic history of the specific animal population (e.g., feralization), selective breeding, the degree to which inter-breeding has taken place with the wild population and the intensity as well as kind of management (like herding, free-range, penning among the others) rather than expecting them to change according to the absence or presence of human control.

The end goal of many domestication studies is the reconstruction of the management regimes, as a result of which greater emphasis is being laid in the recent studies on using stable isotopes whilst identifying the manipulation of diet and mobility caused by humans to contextualize phenotypic changes and differentiating strategies of intensive management by using the composition of sex and age of the animal population that has been slaughtered.

The focus of these differentiating strategies is on the culling of surplus, young males rather than on hunting were more often than not the adult males are targeted (Arbuckle and Atici, 2013; Zeder and Hesse, 2000).

Attempts can now be made to reconstruct the development and origin of systems involved in the management of animals, especially during the Neolithic Near East, by analyzing a series of demographic, isotopic, biometric, and paleogenetic datasets available from various sites across Southwest Asia.

When a human does an act of adapting a particular species and making some modification in the traits of that species. The modification is done so that the species can be clearly distinguished from its native origins. This process is known as domestication. There are evidences of domesticating dogs from wolves over 100,000 YBP (Vila et al., 1997).

It is believed that the process of domestication is much more recent that has begun about 10,000 to 5,000 YBP. It refers to the domestication of animals other than Canines. The process of domestication has started at numerous locations in the world from different periods independently from 10000 to 5000 YBP.

During this period, the population of humans around the world has been estimated to have increased from about 4 million 10,000 YBP to 14 million 5000 YBP. The European and Asian continents has recorded the highest increase in the human population. In these continents, domestication was being broadly adopted for both plants and animals (McEvedy and Jones, 1978).

This was the period that saw the beginning of agriculture. The process of domestication had affected a variety of changing global conditions that affects climate, culture, and society.

Domestication is known to be developed in a wide variety of conditions such as it occurred in both the cultures that are sedentary and nomadic cultures, in agrarian and hunter-gatherer societies. This was before the domestication of plants with the species that is either monogastric or foregut fermentation digestive physiologies.

Even when all the conditions and opportunities were optimal, domestication did not occur in those situations also (Reed, 1977). However, various animal species were tamed during the period. For example, Elephants in India and Hyenas in Egypt.

While only a few species of animals were domesticated. In 1996, there were around 90% of the grazing animals were sheep, goats, or cattle. Those can be traced to just three species-these are *Bos primigenius*, *Ovis Orientalis,* and *Capra aegagrus* (Clutton-Brock, 1981).

For most domesticated animals, it requires almost a period of 100 to 200 years (almost 30 generations) for domestication. However, humans have limited the application of domestication to only a few species. But it is also true that the traditional animal husbandry practice has been conveyed successfully to the human generations.

Communities throughout the world were engaged in the traditional animal husbandry process such as selective breeding practices, using grazing system and spreading the animal manure to the agricultural lands. This practice was adopted by 5000 YBP. There were around 3.3 billion grazing livestock in the year 1996.

There were 91% of domesticated grazers of the total world population of domesticated animals. Those were primarily cattle, sheep, horses, mules, camels, buffaloes, and goats, including pigs and poultry (FAO, 1996). There were a 1% of the annual increase in the world population of grazing livestock such as cattle, sheep and goats since World War II.

However, the increase was simply heterogeneous for the last five decades. The numbers doubled in South America, Africa and Eastern Europe, where the rate of increase was higher than 50% of the global average. In Asia, the increase in the number of sheep has been 77% of the global increase.

During the century, one of the main issues that the continent suffered was overgrazing. Overgrazing has been cited as the main reason for land degradation (Dregne et al., 1991).

During the excavations of the fire pits in the ancient kitchens, animal bones have been found from dinners and such other social gatherings of humans, which show them to have been domesticated. By 8000 before the Common Era (BCE) throughout Asia, sheep and goat were for certain domesticated even though in all probability, animals were domesticated even earlier than this.

Pigs were domesticated around 6500 BCE, in 6700 BCE, it was sheep; in 7000 BCE, it was goats, and as far back as 7700BCE, wheat was widely used and domesticated in Mesopotamia. When Eridu, the first city of Mesopotamia, settled, domesticated animals were being widely used for ploughing, and in another workforce, they have kept as pets as well as a source of food and animal husbandry was being practised extensively.

Over a while, horses became important for usage during the war by various nations, especially for drawing great chariots even though it was around 4000 BCE that humans had begun to tame them.

Subsequently, other animals like tigers, elephants, and lions too started getting deployed during warfare, especially during the Persian campaigns, in resistance to Alexander the Great as a part of the Indian war strategy and the better remembered Punic Wars where they were used against the Romans by Hannibal of Carthage.

The lifestyle of the people changed tremendously in the regions where animals were domesticated. Permanent settlements with pastoral existence slowly started becoming the norm where earlier the main means of subsistence comprised of gathering and hunting so that civilizations not started relying on crops and cattle.

The cultures existing at the time slowly started absorbing animals in their various rituals irrespective of whether they were widespread or basic to the rituals practised once people realized that taming of animals as possible.

Animal imagery was incorporated in the practices and religious icons of various ancient cultures even though the worship of animals in Egypt is more extensively known, especially cats who symbolized Bastet, the goddess of home and hearth.

In places like Rome and Greece, the dog (which is a domesticated animal) was revered where it symbolized security and comfort, and on the other hand, in places like Mesopotamia, the lions (wild animal) of goddess Inanna were revered as it represented the forces of the universe which were untamed.

According to Durant, India during ancient times strongly believed in reincarnation and karma's infinite web so much so that the people felt the souls of humans and animals were the same and continuously moved from one to the other in this never-ending cycle of birth and death as a result of which there was no real gap between men and animals.

For instance, the Hindu god Ganesha has been widely accepted as Shiva's son. Ganesha, or the elephant god, represented the animal nature of humans, whereas his image is considered a lucky charm that helps ward off evil fortune.

In opposition to the chaos prevalent in the world of the untamed or the wild, the order was symbolized by the domesticated animals. After 7000 BCE in Mesopotamia, the number of bones of domesticated goats and sheep showed a steady rise as compared to that of the wild gazelle bones in the refuse dumps that existed outside the cities and towns, which shows that wild game essentially reduced during the period which pertains to those excavations.

Similar patterns have been found to have been prevalent in India, Egypt, and China. Based on the artwork and writings of these cultures, along with the general condition of the bones found, scholars have concluded that goats and sheep were tamed and domesticated.

It is widely presumed that to avoid and escape natural predators, wild goats and sheep would have come closer to human settlements for grazing purposes as those predators would, in turn, have avoided human contact.

Over a period of time, these animals would have been accessible easily for food purposes and could be tamed as well with equal ease. Essentially this process of gradually taming wild animals due to close proximity with humans is believed to have been the reason and means for the domestication of both cats and dogs.

1.5.1. Cats and Dogs

In ancient Egypt, animal husbandry reached its zenith where tamed dogs and cats were treated and cared for as a part of the family with which they dwelled. Herodotus shares that these two animals were so well looked after that their eyebrows would be shaved, and upon the demise of a cat or dog, to mourn the death funeral procession would be taken out.

The feelings and regard that the Egyptians had for their pet animals are evident in the discovery of mummies of dogs and cats in various Egyptian tombs. In 525 BCE, during the well-known Battle of Pelusium,

Egypt was defeated by Cambyses II of Persia, who very strategically drove out the animals loved by Egyptians before their frontline and had the shields of his soldiers painted with an image of goddess Bastet (the great cat goddess). Most of the Egyptians were massacred as they fled the battlefields without putting up a fight for fear of offending their gods if they, by any chance, hurt any of those animals.

The conquering of Egypt on these lines made Cambyses of Persia contemptuous of the Egyptians, who, according to him, preferred to surrender their freedom rather than harm their animals. During his triumphal march post his victory in the battle; he went a step ahead by hurling cats onto the Egyptians' faces.

1.5.2. Sheep

Amongst all the livestock to have been domesticated, sheep was probably the earliest ones to have been tamed. Genetic studies have gone to show the Asiatic mouflon (Ovis Orientalis) was the origin for all domestic sheep (*Ovis aries*) (Chessa et al., 2009).

The assemblage on archaeo-faunal data of the late Pleistocene shows 50–90% of total hunting was for goat and sheep, especially in the region between the Zagros and the western Taurus mountains.

Culling patterns (during the early tenth millennium) that were focused on the juvenile (c. 50% of them were slaughtered before the age of two) have been observed in the Shanidar Cave (level B1) and the open site close by of ZawiChemi Shanidar through dating done in the northern Zagros.

This has been recognized by Perkins (1964) as one of the earliest evidence showing intensive management. More recently (2008a, 2011), the culling of young ZawiChemi sheep has been confirmed by Zeder.

At the same time, a shift in the various hunting strategies has been acknowledged by Zeder and the culling of young animals (aged between 2 and 3) has also been evident in the contemporary sites of Körtik Tepe and Hallan Çemi (Arbuckle and Özkaya, 2007; Redding, 2005; Zeder, 2012a).

A trend towards domestication of wild sheep seems evident from the findings of round architecture at these sites which indicates an increase in the management as well as intensive interaction of humans with the population of wild sheep which then showed phenotypical changes and some level of sedentary lifestyle in the wild sheep (c. 50% NISP (number of identified specimens)).

Observing the remains at the nearby site of Göbekli Tepe it can safely be said that at times around these early settlements' animal were not only captured and kept there but were also culled from the population that was free-ranging even though there is no clarity about the exact relationship that existed between humans and animals around this tenth millennium (cal BC) period.

Even though it is difficult to identify the exact nature of experiments conducted for domesticating the wild mouflon population due to the easy accessibility to this species, the ephemeral and small-scale management followed at that time, it can be said that eastern Turkey, NW Iran, and NE Iraq played an important role in the development of intensive techniques for managing them.

This is evident from the fact that there was a focus on the exploitation of sheep in the upper Tigris drainage's eastern parts (Meadows et al., 2007). At Cafer Höyük, the 8th and 9th millennium cal Before Christ (BC) site located in the Upper Euphrates highlands too, it has been difficult to identify the exact management strategies that were followed in assemblages.

According to Helmer (2008), along with the hunting of the mouflon, the morphologically wild sheep was being herded. It was only in the 8th-millennium deposits that this parallel strategy of herding and hunting was identified (phases I–IV), but it is quite possible that this strategy existed even during the early stages of these settlements dating back to the ninth millennium (phases XXIII–IX) whereby the wide area of the Northern Zagros and Anti-Taurus had different forms of sheep management.

Outside the region where intensive hunting of mouflon took place, in the lowlands, especially around the south-eastern parts of Turkey, the evidence for sheep management is far clearer for the mid-ninth millennium cal BC period. In the early ninth and tenth millennia, wild sheep, more particularly, large rams were hunted once in a while at Çayönü in the earliest phases (Round and Grill phases) (8% NISP) (Hongo et al., 2005).

By the mid to late ninth millennium during the Channeled Building phases (8400–8200 Cal BC) even though small scale but intensive management of sheep took place which is evident from the fact that smaller sized sheep appeared around this period and culling of female sheep became more prominent along with a steady increase in the overall sheep population during the Cobble-paved phase.

Angora, Alpine, French, Nubian, Saanen, and Toggenburg are the main exotic sheep breeds, and they primarily feed on wild plants including green grass. Minerals, proteins, and food rich in vitamin are given to sheep when they are bred for a specific purpose. Their food mainly consists of:

- Leguminous fodder: moong, urad, berseem, etc;
- Grains: Barley, jowar, maize, and oats;
- Oil cakes: Sesame cake which is rich in proteins and groundnut cakes;
- Common salt, lime, and sterilized bone meal which is a rich source of mineral salts.

1.5.3. Goat

Asiatic bezoar (Capra aegagrus) has been found through various genetic studies to be the ancestors of the domestic goats (*Capra hircus*) (Luikart et al., 2001; Naderi, 2007). Evidence is available to show that in the ninth-millennium cal BC in various regions like the throughout the entire Fertile Crescent in the upland regions' goats were exploited extensively, unlike the mouflon, and their management was quite intensive.

Further evidence has been found for Epipaleolithic and the tenth-millennium cal BC (PPNA (pre-pottery Neolithic A)) that shows the hunting of goats in the central and southern Taurus (Arbuckle and Erek, 2012; Atici, 2011), which was practised in the further south like in Lebanon.

Furthermore, the record systems of intensive exploitation of bezoar during this period have been found at Nachcharini (65% NISP) and Ksar Akil (c. 30% NISP). (Garrard et al., 2003; Kersten, 1989; Kuhn et al., 2009).

Right from the Younger Dryas to the eighth-millennium cal BC the animal economy was replete with the hunting of bezoar in southwestern Syria (Helmer and Gourichon, 2008; Ibanez et al., 2010). During the early phase (52% NISP), namely from the mid to the late ninth-millennium cal BC morphologically wild goats were the most abundant species at Aswad, which was at a distance of about 20 kilometres from the habitat of bezoar.

It seems that the movement of the goats in the lowland regions was intentional, and these animals were available in the lake basin area in large numbers even though for this early phase, there is not much evidence of a zoo-archaeological kind.

Goats were intensively managed during the late ninth millennium (middle phases) in southwestern Asia, as is evident from the presence of torsion on male hormones and an increased prominence given to the culling of young males.

At Aswad, as well as the site of Ghoraife which was nearby morphologically domestic goats became an essential and important part of the economies accounting for 33% and 42%, respectively, by the early part of the 8th millennium (Ducos, 1993).

During the late 9th and early part of the 8th- millennium, in the Jordon Valley which was further south, the economies focused majorly on goats. From the period starting in the late 10th millennium (PPNA) (3% NISP), the number of goats increased extensively at Jericho, which carried on through the 9th and till the 8th-millennium (PPNB), representing about 52% of the total animal population (Clutton-Brock, 1979).

During the early and middle parts of the eighth-millennium goats were the most prominent species (65% NISP) at Ain Ghazal as is evident from the presence of various factors like the presence of certain morphological changes which included a higher variation on the morphology of horncore, a reduction in the size of the goats and culling of young males (Wasse, 2002).

Open fields with plenty of greens are the feeding areas for goats. In addition to that, they can be fed grain products and cereal whereas at times a balanced feed comprising of a mixture of grains that have been crushed like jowar and maize, certain cereals, four to five kilograms of fodder, steamed bone meal, linseed oil meal and groundnut.

1.5.4. Cattle

During the Pleistocene and early Holocene period in most of the southwestern parts of Asia except the aridest ones had the domestic taurine cattle (Bos Taurus), which has its origins in the aurochs (Bos primigenius) (Bradley et al., 1998; Uerpmann, 1987).

Zooarchaeological evidence shows that even though intensive cattle management was practised only later in the regions neighbouring the upper Tigris and the Syrian Euphrates, genetic, and faunal research have both gone to show during that these regions practicsed intensive management of cattle (Bollongino et al., 2012; Helmer et al., 2005).

At the site of Mureybet, along the Syrian Euphrates, there was a major amount of hunting of the aurochs (Gourichon and Helmer, 2008). Of the total mammals amidst the varied species of animals present here during the ninth millennium, 30% were morphologically wild cattle (levels IVA and B) and the number was around 18% from Jerf El Ahmar, which is nearby.

Furthermore, 18% of the NISP in the north, namely at Gobekli Tepe consisted of aurochs (Peters et al., 2005). Early ninth-century evidence exists especially about Djabe where Helmer and his colleagues (2005) found a reduced sexual dimorphism which leads to the fact that intensive management caused morphological changes in cattle.

At Halula, dating the earliest layers from the early 8th millennium, as well as mid-8th millennium taken on the Turkish Euphrates from both Gritille and Teleilat, shed evidence on the striking reduction in the size of the body (Ilgezdi, 2008; Monahan, 2000).

Even though the hunting of aurochs continued at Cafer (between 8300 and 7500 BC), which was further north, cattle herding has been evident in the various pathological factors existing along with a little reduction in dimorphism as had happened at Djabe (Helmer, 2008).

A few of the varieties that give a high yield include Jersey, Holstein, Swiss, Friesian, etc. For about 300 days, the milk production period lasts which is essentially the lactation period or the time gap between two pregnancies. Even though the grass is the main source of food for buffaloes and cows, it is not sufficient to provide the entire nourishment.

Fibrous food loaded with fibre provides them roughage which can be catered for through hay fodder, peas, cereals like jowar, and maize, etc., and leguminous plants like soybean.

Full cream milk is a dairy product that is obtained from animals. Toned milk is the result of the removal of cream from this milk which is devoid of all fat and is also called skimmed milk. The fat content of the different milk products distinguishes them into the following categories:

1. **Cream:** The fat content in cream is between 10 and 70% and is obtained by skimming off the fat which comes on the top of the milk after it has been churned.
2. **Butter Milk:** On the removal of butter from the milk, the liquid that is left is called buttermilk.
3. **Ghee:** Water evaporates totally when butter is heated, leaving behind 100% fat called ghee.

4. **Curd:** Bacterial activity helps to curdle milk and results in a healthy curd.
5. **Powdered Milk:** Milk in its powdered form wherein it is essentially evaporated to dryness forms powder milk which has a longer shelf life.
6. **Condensed Milk:** By removing water milk is concentrated either with or without the addition of sugar forming condensed milk.
7. **Khoya:** The bulk of the milk is reduced to around 70–75% by evaporation of its water content and obtaining this desiccated form of milk product.
8. **Cheese:** This is essentially the product obtained by the coagulation of fat and water with casein, the milk protein.
9. **Cattle Dung:** Used mostly in the villages of India, cow dung is used as a fuel in the form of dung cakes. Biogas is also made from cattle dung whereby the farmers use this biogas, and the remaining part is used as a rich source of manure.

Animal excreta which mainly comprises buffalo and cow dung is fed into an airtight biogas chamber and treated with some anaerobic activity in biogas. The methane (CH_4) gas thus produced through the decomposition of excreta is used in cooking as a gas that produces no smoke. Another use of this smokeless gas is for lighting and rich manure is obtained from the solid residue that is left over.

1.5.5. Pigs

Throughout southwestern Asia in the river valleys, the wild boar (Sus scrofa) existed, which gave form to the domestic pigs. The history of pig husbandry is quite complex, and genetic studies dealing with this matter show that some of the earliest regions for the domestication of pigs were in the Near East (Larson, 2005).

The environmental requirements along with the mobility and diet of these non-ruminant omnivores are quite different from the other domesticated livestock of the Neolithic age.

Stalling and foddering as well as a large amount of free-range management of pigs shows a variety of relationships that may have been characteristic to the people-swine relationship due to their intensive control which varied dramatically from the management of sheep, cattle, and goats (Albarella et al., 2007; Hadjikoumis, 2012).

It becomes extremely challenging to understand the history involved in the management of pigs due to the relationship they shared with humans as also the immense amount of flexibility in the population of pigs (Figure 1.4) (Rowley-Conwy et al., 2012).

Figure 1.4. Pig management.

Source: Image by Farmers Weekly.

During the early Neolithic age, there was an immense amount of exploitation of suits in the island of Cyprus as well as in the Euphrates and upper Tigris basins. In the late Pleistocene age during the 11th and through the ninth millennia primarily boars were hunted, which called for their transportation to Cyprus.

This has been clearly indicated in some of the recent research studies (Vigne et al., 2009, 2012). However, according to Vigne, it was only during the early part of the 8th-millennium that pig husbandry was practised extensively even though evidence shows that the bear population was manipulated even before this (Vigne et al., 2011).

Some of the pig breeds include India Middle white Yorkshire England, Large white Yorkshire England, Hampshire Tamworth England Bershile England, Essex saddleback, and India Landrace Denmark.

1.6. CONCLUSION

At the end, it is concluded that animals need effective care and management as they provide us with a huge variety of food products having high nutritional values. Animals are commercially to meet the high demands of food. Dairy products from different animals like goats, cows, buffaloes are considered rich sources of protein.

As they provide us milk, they are often called milch. Another set of animals that provide high-nutrient food are ducks, hen, goose, etc. Animals like a duck, chicken, pigs, goat, etc., are bred for meat. The care and breeding management of animals are usually observed under the department of animal husbandry. It can be said that animal husbandry is a large business.

The animals are bred, cared for, reared, and sheltered in a farm or region, that are is mainly built for them. Animal husbandry further involves poultry, milk farms, apiculture (bee agriculture), aquaculture, etc.

REFERENCES

1. (Author name not found with text). *Live Stock Production and Management*. [eBook] Tamil Nadu, India: Tamil Nadu University of Agriculture. Available at: https://drive.google.com/file/d/1SXEi1HiQ4UnWkAMO4aMg0QdvICO-2kZC/view (accessed on 15 March 2021).
2. Albarella, U., Johnstone, C., & Vickers, K., (2007). The development of animal husbandry from the Late Iron Age to the end of the Roman period: a case study from South-East Britain. *Journal of Archaeological Science,* 35(7):1828-1848. DOI:10.1016/j.jas.2007.11.016.
3. Arbuckle, B. S., and Atici, L., (2013). Initial diversity in sheep and goat management in Neolithic Southwestern Asia. *Levant*, 45:219-235.
4. Arbuckle, B., & Erek, C., (2012). Late Epipaleolithic hunters of the Central Taurus: faunal remains from Direkli Cave, Kahramanmaras, Turkey. *International Journal of Osteoarchaeology*, 22: 694-707.
5. Arbuckle, B.S., & Özkaya, V., (2007). Animal exploitation at Körtik Tepe: An early Aceramic Neolithic site in southeastern Turkey. *Paléorient,* 32: 198-211.
6. Arbuckle, B., (2014). *Pace and Process in the Origins of Animal Husbandry* (8th edn., pp. 53–81). Neolithic Southwest Asia.
7. Asouti, E., & Fuller, D.Q., (2012). From foraging to farming in the southern Levant: The development of Epipalaeolithic and Pre-Pottery Neolithic plant management strategies. *Vegetation History and Archaeobotany,* 21(2):149-162. DOI:**10.1007/s00334-011-0332-0**.
8. Atici, L. 2011. Before the Revolution: Epipaleolithic Subsistence in the Western Taurus Mountains, Turkey. British Archaeological Reports International Monograph Series 2251. Oxford: Archaeopress.
9. Banning, E.B., (1998). The Neolithic Period: Triumphs of Architecture, Agriculture, and Art. *Near Eastern Archaeology*, 61(4):188-237. DOI:**10.2307/3210656**.
10. Bellwood, P., & Dizon, E., (2005). The Batanes Archaeological Project and the "Out Of Taiwan" hypothesis for Austronesian dispersal. *Journal of Austronesian Studies*, 1:1–33. Taitung, Taiwan.
11. Bhat, P., & Yadav, M., (2018). *Animal Husbandry*. Jodhpur-India, New Delhi-India: Scientific Publishers (India).
12. Bokonyi, S., (1969). Archaeological problems and methods of recognizing animal domestication. In: Ucko, P.J.- Dimbleby, G.W.

(eds.), The domestication and exploitation of plants and animals, London, 221-229.

13. Bollongino, R., Burger, J., Powell, A., Mashkour, M., Vigne, J. D., & Thomas, M. G., (2012). Modern taurine cattle descended from small number of near-eastern founders. *Molecular biology and evolution*, 29(9): 2101–2104. https://doi.org/10.1093/molbev/mss092.
14. Bradley, D.C., Chang, G.C., & Andersen, R.A., (1998). Encoding of three-dimensional structure-from-motion by primate area MT neurons. *Nature,* 392:714–717.
15. Cauvin, J., (2000). *The Birth of the Gods and the Origins of Agriculture (New Studies in Archaeology)*. Cambridge University press, Cambridge, Pp-259.
16. Chessa, B., Pereira, F., Arnaud, F., Amorim, A., Goyache, F., Mainland, I., Kao, R. R., Pemberton, J. M., Beraldi, D., Stear, M. J., Alberti, A., Pittau, M., Iannuzzi, L., Banabazi, M. H., Kazwala, R. R., Zhang, Y. P., Arranz, J. J., Ali, B. A., Wang, Z., Uzun, M., … Palmarini, M. (2009). Revealing the history of sheep domestication using retrovirus integrations. *Science (New York, N.Y.)*, 324(5926), 532–536. https://doi.org/10.1126/science.1170587.
17. Clutton-Brock, T. H., & Harvey, P. H., (1979).Comparison and adaptation. Proceedings of the *Royal Society* of *London*. Series *B*, 205: 547–565 (doi:10.1098/rspb.1979.0084)
18. Clutton-Brock, J.. (1981). *Domesticated animals from early times*. London, British Museum (Natural History) and William Heinemann Ltd. pp.208.
19. Cunliffe, A., (2008). *Organization theory*. 10.4135/9781446214770.
20. Dregne, H.E., Kassas, M., & Rozanov, (1991). A new assessment of the world status of desertification. *Desertification Control Bulletin,* 19: 6-18.
21. Ducos, P., (1978). Domestication" defined and methodological approaches to its recognitionin faunal assemblages. Pp. 49–52 in RH Meadow and MA Zeder (eds.) Approaches toFaunal Analysis in the Middle East.Peabody Museum Bulletin 2. Cambridge, MA: Harvard University Press.
22. Ducos, P., (1993). Some remarks about *Ovis*, *Capra* and *Gazella* remains from two PPNB sites from Damascene, Syria, Tell Aswad and Ghorife. In:H.Buitenhuis and A.T. Clason (eds.), *Arachaeozoology of the Near East* vol. I:37-42. Leiden, Universal book services.

23. Dudeja, P., Singh, A., & Kaur, S., (2016). *Good Animal Husbandry Practices*. New Delhi: CBS Publishers and Distributors Pvt. Ltd.
24. FAO (1996). Food, Agriculture and Food Security: Developments since the World Food Conference and Prospects, Technical Background Document No 1 for the World Food Summit, Rome.
25. Fuller, C., Michie, S., Savage, J., McAteer, J., Besser, S., Charlett, A., et al. (2012). The Feedback Intervention Trial (FIT) — Improving Hand-Hygiene Compliance in UK Healthcare Workers: A Stepped Wedge Cluster Randomised Controlled Trial. *PLoS ONE,* 7(10): e41617. https://doi.org/10.1371/journal.pone.0041617.
26. Garrard, S.M., Capaldo, C.T., Gao, L., Rosen, M.K., Macara, I.G., & Tomchick, D.R., (2003). Structure of Cdc42 in a complex with the GTPase-binding domain of the cell polarity protein, Par6. *EMBO J.*22: 1125–1133.
27. Gourichon, L., & Helmer, D., (2008).Etude archeozoologique de Mureybet, pp. 115e228.In: Ibanez, J.J. (Ed.), Le site neolithique de Tell Mureybet (Syrie du Nord), OxfordBAR International Series 1843.
28. Guilaine, J., & Manen, C., (2007). From the Mesolithic to Early Neolithic in the Western Mediterranean. *Proceedings of the British Academy*, 144:21-51. DOI:10.5871/bacad/9780197264140.003.0003.
29. Hadjikoumis, A., (2012). Traditional pig herding practices in southwest Iberia: Questions of scale and zooarchaeological implications. Journal of Anthropological Archaeology 31 (3), 353-364.
30. Hammer, K., (1984). Das Domestikationssyndrom. *Die Kulturpflanze* 32: 11–34. https://doi.org/10.1007/BF02098682.
31. Helmer, D., (2008). Révision de la faune de Cafer Höyük (Malatya Turquie): Apports des méthodes de l'analyse des mélanges et de l'analyse de kernel à la mise en évidence de la domestication. In Vila, E., Gourichon, L., Choyke, A. M., and Buitenhuis, H. (eds.), *Archaeozoology of the Near East VIII*, Maison de l'Orient et de la Mediterraneé, Lyon, pp. 169–195.
32. Helmer, D., & Gourichon, L., (2008). 'Premières données sur les modalités de subsistance dans les niveaux récents de Tell Aswad (Damascène, Syrie) – fouilles 2001-2005', in E. Vila, L. Gourichon, H. Buitenhuis & A. Choyke (eds), Archaeozoology of the Southwest Asia and Adjacent Areas VIII, 119–151. Travaux de la Maison de l'Orient 49, Lyon.

33. Helmer, D., Gourichon, L., Monchot, H., Peters, J., & Saña, S. M., (2005). Identifying early domestic cattle from Pre-Pottery Neolithic sites on the Midddle Euphratesusing sexual dimorphism. In: Vigne J.-D, Helmer D, Peters J, editors. The first steps of animal domestication: new archaeozoological approaches. Oxbow Books; Oxford, UK: pp. 86–95.

34. Hole, F., Kent, V. F., & James, A. N., (1969). Prehistory and Human Ecology of the DehLuran Plain. Memoir, Museum of AnthropologyVol. 1. Ann Arbor, MI: University of Michigan.

35. Hongo, H., Meadow, R.H., Oksuz, B., & Ilgezdi, G., (2005). Sheep and Goa tRemains from Çayönü Tepesi, South eastern Anatolia, in Buttenhuis, H, Choyke, A.M., Martin, L., Bartosiewicz, L & Mashkour, M. (eds), *Archaeozoology of the Near East VI. Proceedings of the sixth international symposium on the archaeozoology of southwestern Asia and adjacent areas*. ARC-Publicaties,Groningen:113-124.

36. Ibanez, I., Primack, R.B., Miller-Rushing, A.J., Ellwood, E., Higuchi, H., Lee, S.D., Kobori, H., & Silander, J.A., (2010). Forecasting phenology under global warming. *Philosophical Transactions of the Royal Society B – Biological Sciences* 365: 3247– 3260.

37. Ilgezdi, G., 2008. The Domestication Process in Southeastern Turkey: the Evidenceof Mezraa-Teleilat. PhD Dissertation. Eberhard-Karls-Universita€tTu€bingen.

38. Kersten, A.M.P., (1989).The Epipalaeolithic ungulate remains from Ksar Akil: Somepreliminary results [in:] "People and culture in change. Proceedings of the second symposium on Upper Paleolithic, Mesolithic and Neolithic populations of Europeand the Mediterranean Basin", Hershkovitz I. (ed.), BAR International Series 508, Oxford: Archaeopress, pp. 183-198.

39. Kühn, H., Hierlmeier, T., Merl, J., Jakob, S., Aguissa-Touré, A. H., Milkereit, P., & Tschochner, H., (2009). The Noc-domain containing C-terminus of Noc4p mediates both formation of the Noc4p-Nop14p submodule and its incorporation into the SSU processome. *PloS one*, *4*(12), e8370. https://doi.org/10.1371/journal.pone.0008370.

40. Larson, G., (2005). Worldwide pylogeography of wild boar reveals multiple centers of pig domestication. *Science*, 307:1618–1621.

41. Luikart, G., Gielly, L., Excoffier, L., Vigne, J.D., Bouvet, J., & Taberlet, P., (2001). Multiple maternal origins and weak phylogeographic structure in domestic goats. Proc Natl Acad Sci USA. 98:5927–5932.

42. Mark, J. J, (2010). *Animal Husbandry*. [Online] World History Encyclopedia. Available at: https://www.ancient.eu/Animal_ Husbandry/ (accessed on 15 March 2021).

43. Makarewicz, C., & Tuross, N., (2012). Finding Fodder and Tracking Transhumance: Isotopic Detection of Goat Domestication Processes in the Near East. *Current Anthropology*, 53(4):495-505. DOI:**10.1086/665829**.

44. McEvedy, C., & Jones, R., (1978). *Atlas of World Population History*. Penguin Books Ltd.; New York, NY.

45. Meadow, C.T., Cerny, B.A., Borgman, C.L. & Case, D.O. (1989). Online access to knowledge: System design. J. Am. Soc. Inf. Sci., 40: 86-98. https://doi.org/10.1002/(SICI)1097-4571(198903)40:2<86::AID-ASI3>3.0.CO;2-K.

46. Meadows, J. R., Cemal, I., Karaca, O., Gootwine, E., & Kijas, J. W., (2007). Five ovine mitochondrial lineages identified from sheep breeds of the near East. *Genetics*, 175(3): 1371–1379. https://doi.org/10.1534/genetics.106.068353.

47. Monahan, B.H., (2000). The organization of domestication at Gritille, A Pre-Pottery Neolithic B Site in Southeastern Turkey. Ph.D. Dissertation, Northwestern University.

48. More, H., (2019). *Animal Husbandry: Its Objectives and Products Obtained*. [Online] The fact factor. Available at: https://thefactfactor.com/facts/pure_science/biology/animal-husbandry/2294/#:~:text=Objectives%20of%20Animal%20Husbandry%3A&text=To%20do%20proper%20management%20of,increase%20the%20production%20of%20milk; https://thefactfactor.com/facts/pure_science/biology/animal-husbandry/2294/ (accessed on 15 March 2021).

49. Naderi, S., Rezaei, H. R., Taberlet, P., Zundel, S., Rafat, S. A., Naghash, H. R., el-Barody, M. A., Ertugrul, O., Pompanon, F., & Econogene, C., (2007). Large-scale mitochondrial DNA analysis of the domestic goat reveals six haplogroups with high diversity. *PloS one*, *2*(10), e1012. https://doi.org/10.1371/journal.pone.0001012.

50. Perkins, D. (1964). Prehistoric fauna from Shanidar, Iraq, *Science*, 144:1565-1566.

51. Redding, G., (2005). 'The Thick Description and Comparison of Societal Systems of Cap-italism',*Journal of International Business Studies*,36: 123 – 155.

52. Reed, C. A., (ed.). (1977). *Origins of Agriculture*. The Hague: Mouton. (This is an excellent collection. Reed's own discussion and conclusions are an especially good synthesis.)
53. Rowley-Conwy, P., Albarella, U., & Dobney, K., (2012). Distinguishing Wild Boar from Domestic Pigs in Prehistory: A Review of Approaches and Recent Results. *Journal of World Prehistory, 25*(1), 1-44. Retrieved May 23, 2021, from http://www.jstor.org/stable/41429632.
54. Russell, A.L., (2010). Retracing the steppes: a zooarchaeological analysis of changing subsistence patterns in the late Neolithic at Tell Sabi Abyad, northern Syria, c. 6900 to 5900 BC. PhD. Thesis, Faculty of Archeology , Leiden University, Leiden.
55. Sikkim-ahvs.gov.in. (2011). *Animal Husbandry Livestock, Fisheries and Veterinary Services*. [Online] Available at: http://www.sikkim-ahvs.gov.in/main_objectives.html (accessed on 15 March 2021).
56. Uerpmann, H.P., (1987), *The Ancient Distribution of Ungulate Mammals in the Middle East*, volume 27 of the Tübinger Atlas des Vorderen Orients (TAVO), Series A.
57. Vigne, J.D., Carrére, I., Guilaine, J., & LeBrun, A., (2003). Unstable status of early domestic ungulates in the Near East. in The Neolithic of Cyprus, eds Guilaine J, LeBrun A (École Français d'Athèns, Athens), 239–251, (original in French) Bull Corr Héllenique, Suppl 43.
58. Vigne, J-D., Antoine, Z., Jean-François, S., François, P., Guilaine, J., & Simmons, A., (2009). Pre-Neolithic wild boar management and introduction to Cyprus more than 11,400 years ago. *Proceedings of the National Academy of Sciences of the USA,* 106(38):16131–16138.
59. Vigne, J., Carrère, I., Briois, F., & Guilaine, J. (2011). The Early Process of Mammal Domestication in the Near East: New Evidence from the Pre-Neolithic and Pre-Pottery Neolithic in Cyprus. *Current Anthropology, 52*(S4), S255-S271. doi:10.1086/659306.
60. Vigne, J-D., Briois, F., Zazzo, A., Willcox, G., Cucchi, T. Stéphanie, T., Isabelle, C., Yodrik, F., Régis, T., Chloé, M., Christophe, M., Clothilde, C., & Guilaine, J., (2012). First wave of cultivators spread to Cyprus at least 10,600 y ago. *Proc. Natl. Acad. Sci. USA* 109:8445–49.
61. Vila, C., Savolainen, P., Maldonado, J.E., Amorim, I.R., Rice, J.E., Honeycutt, R.L., Crandall, K.A., Lundeberg, J., & Wayne, R.K., (1997). Multiple and ancient origins of the domestic dog. *Science*. 276: 1687-1689.

62. Wasse, A., (2002). Final results of an analysis of the sheep and goatbones fron Ain Ghazal, Jordon, *Levant*, 34:59-82.

63. Zeder, M.A., (2006). Central questions in the domestication of plants and animals. *Evolutionary Anthropology*, 15: 105-117. https://doi.org/10.1002/evan.20101.

64. Zeder, M.A., (2008b). Domestication and early agriculture in the Mediterranean Basin: Origins, diffusion, and impact. Proceedings of the National Academy of Sciences Aug 2008, 105 (33) 11597-11604; DOI: 10.1073/pnas.0801317105.

65. Zeder, M., (2011). The origins of agriculture in the Near East. *Current Anthropology,* 52: 221-35.

66. Zeder, M., (2012a). The broad spectrum revolution at 40: resource diversity, intensification, and an alternative to optimal foraging explanation. *Journal of Anthropological Archaeology* ,31: 241-264.

67. Zeder, M., (2012b). The Domestication of animals. *Journal of Anthropological Research, 68*(2), 161-190. Retrieved May 23, 2021, from http://www.jstor.org/stable/23264664.

68. Zeder, M., & Hesse, B., (2000). The Initial Domestication of Goats (Capra hircus) in the Zagros Mountains 10,000 Years Ago. *Science,* 287(5461):2254-2257. DOI:10.1126/science.287.5461.2254.

Chapter 2

Technology and Science behind Animal Husbandry

CONTENTS

Technology and science play a very important role in the overall development of animal husbandry. In this chapter, several technologies that help in overall animal health are discussed. It also highlighted biotechnology used in the livestock industry. In addition, it also thrown some light on technological innovations in animal production, which is related to environmental sustainability. In the end, it explains the application of information technology in animal health.

2.1. INTRODUCTION

It is well known that all of us are aware of the term "farm-to-table." We often set out on sunny Saturday mornings for brunch at trendy restaurants and cafés, but while we gulp down our mimosas, we generally overlook the undertaking needed to bring our preferred dishes from farms to kitchens.

Farmers of the current era are facing challenges from basic infrastructure to connectivity, rising demands for animal proteins to food spoilage, and uncertainty that affect the overall health of animals.

Technology has set up its footprints in every field, and AgTech is no exception. Thus, it is essential to get back to basic building blocks and congregate viewpoints from those leading the industry and working hard to address these issues.

The conventional livestock industry is a sector that is extensively unnoticed and under-serviced, though it is the most imperative one. Livestock provides much needed renewable as well as natural resources that are vital for survival on this earth.

So, why is the process to implement advanced technology in this industry is so slow? Well, the exact and precise answer, money. The economics of the livestock industry changes from season to season implies it is not as profitable always as it looks like. With limited investments, comes fewer acceptances of new technology.

There is a need on the part of investors need to recognize the market dynamics that farmers face. "We can't view farming in the Silicon Valley mind frame since farms don't operate in the same manner," says Amado Guloy, CEO, and Founder of Rex Animal Health.

Internet of things (IoT), IoT has taken off, and investments in IoT will likely to further in the future, yet the majority of the investors do not recognize that big data is often unreachable to farmers because of lack of internet connectivity and infrastructure facility.

Several companies are actively working in this segment to address some of these industry's pain points. For instance, some technology equipment has the potential to monitors cows 24/7, 365 days a year, examining or interpreting their well-being, performance, and productivity.

Using artificial intelligence and computer vision, these technologies decode visual information into actionable data. In seconds, this imaging technology recognizes and remembers individual animals by their distinctive features.

Apart from these, there are several record-keeping software that record and store data related to performance, daily activities, health-related information, climatic conditions and their subsequent impact on animals, and genetic data to provide a clear understanding for farmers to predict and prevent disease in herds, that in turn can enhance the overall yields.

Livestock management has traditionally been considered as the method to manage or run the poultry business, cattle ranches, dairy farms, or other livestock-related agribusinesses. Livestock managers are generally concerned with keeping accurate financial records, supervise workers, routine activities, and ensure proper care and feeding of animals.

Although, some current trends have proven that technology is transforming the world of livestock management. Recent developments have made considerable improvements in the industry that make managing and tracking livestock much easier and data-driven. This technology can come in the form of genetics, digital technology, nutritional technologies, and many more (Figure 2.1).

Figure 2.1. Traditional methods usage in animal husbandry.

Source: *Image by Wikipedia.*

Now, there are sensors available in the market intended at monitoring the real-time quality of milk, health conditions of cattle, and pregnancy hormones. Apart from this, virtual fences are in place that can transfer animals wearing a sensor to be moved remotely from one area of a pasture to another.

Even some robots exist, and some are in the development phase in this industry, which is aimed at addressing the challenges of labour shortages on conventional livestock farms. 12% of dairy farms are presently using robots and is projected to grow to 20% in the next 5 years.

Presently, in the digital age is connected, including agriculture and farming. Livestock technology can improve or augment the overall production capacity, management or welfare of livestock and animals. The new concept that is in trend nowadays is 'connected cow.'

It is a result of more and more dairy herds being equipped or installed with sensors to monitor productivity and health-related information. Placing individual wearable sensors on cattle can keep an eye on daily activity and health-related concerns while at the same time providing data-driven insights for the entire herd.

All this data being produced with the help of this technology can be turned into meaningful, actionable insights where producers can look easily and quickly to take prompt management decisions.

2.2. TWO KEY TECHNOLOGIES HELPING AGRICULTURE AND ANIMAL HEALTH

It is generally seen that within the past few years, farms have been significantly developing with respect to the latest technologies and new applications.

According to a recent article, the farming industry is sitting on the edge of the main shake-up in what's being referred to as "Agriculture 4.0."

A forecast which has been projected by the Institute of Biological, Environmental, and Rural Sciences (IBERS) report it that "[envisions] a future where farmers use mobile command centers to oversee robots, power plants, and closed-loop recycling systems. " Imagine if mobile phones were able to control farms.

The IoT is the fuel behind this revolution within the farming community, on a scale that can be compared with the industry's 20th Century mechanization. This recent transition in technology is being signalled as a net positive for the whole society and reformist farmers, affecting the overall production of food, global health, and climate change for the better.

"This raft of technologies could have an enormously positive effect on rural livelihoods, economies of countries, and the potential of the food supply chain to address the greater challenges of our age," says the Institute of Biological, Environmental, and Rural Sciences (IBERS).

It is important to note that IoT doesn't halt with cows. The majority of the farmers are maximizing the use of IoT devices to keep an eye on the ambient conditions in the soil and air-water to organize resources such as watering and pesticides practices.

Agriculture 4.0 is basically concerned with bringing information technology to the farm and ensures its practical usage as we as create a food supply chain that will be connected to all the major components. So, what are several transitions that one is already started experiencing to see automate the way farmers do things?

2.2.1. Rethinking RFID

Radio frequency identification (RFID) or also known as electronic identification (EID) tags are not the newer technology. Although, these tags would become the major part of their daily usage object as with the help of this, farmers will be able to track and identify the stock, automating the data collection process and information out in the field (Figure 2.2).

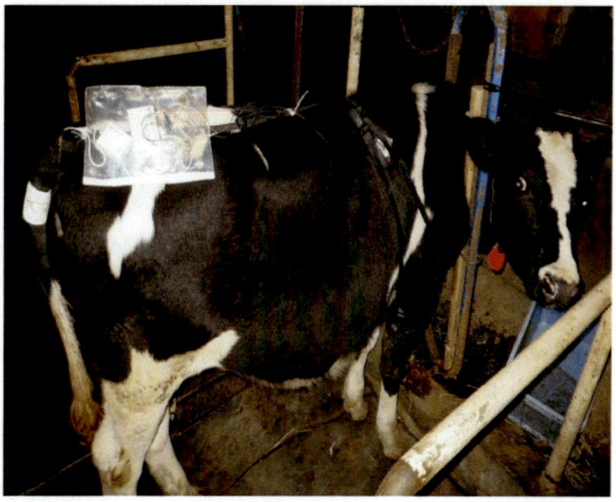

Figure 2.2. Installing RFID on cows.

Source: *Image by Wikipedia.*

It is generally seen that tagging permits animals to be identified, tracked, and located, with the help of radio frequency signals. Over a certain time period, for instance, an animal's weight history can be discovered back as either "positive" or "negative" and, from there, resources can be transferred in a way to upsurge the overall livestock health and profits.

RFID tags and wand readers have is already considered as one of the best ways and has proved their success to track animals in New Zealand after passing the National Animal Identification and Tracing (NAIT) Act of 2012.

In addition, it is also worth noticing that this technology eliminates human error from the equation that may arise while recording stock numbers. RFID tags are still comprehensible even when covered in excrement and dirt, offering a safe distance in which to be working around the stock as it is possible to read the tags from a distance of about 11 to 35 inches (30 to 90 cm).

Radio-frequency identification using electromagnetic fields to automatically identify and track tags attached to objects.

Depending on the type of RFID reader, these can range from anywhere between $350 and $3,500. The majority of the RFID taggers come with a mobile app that permits farmers to transmits the data straight away to NAIT, bringing the tracking full-circle and saving time on documentation.

It has been observed that there are a lot of companies that have already incorporated NAIT functionality into their farming management tools, such as weigh scale software, making more effective and efficient animal management.

It is important to note that it is not only milking cows that would get benefit from this technology. This could also be useful for the pig as well as deer farms through the recording of weight, milk production, treatments, and breeding information. To name just a few!

- **Cow-Sidering Conditions:** It is generally seen that weighing of herd twice in a day when cows exit after milking provides would be really helpful as it provides a more precise reading by averaging the results as well as graphing changes in body weight for each cow. Graphing and Documenting of this data generate a trend line that can be tracked easily and read so labours can react to transition rapidly.

Recent advancements in scales provides a significant role in providing accurate weight for each cow, in addition to balance weight trends with the stage of lactation, daily feed allocation, and milk yield. Combining milk and animal weight data is very crucial in enhancing animal health and tracking transitions in body condition.

2.2.2. More than What "Meats" the Eye: Thermal Imaging (TI)

Thermal imaging (TI) isn't a new concept either, though it is creating or finding its space in animal farms. What if this meant you could identify disease and injuries before these become evident to the naked eye? What if one could simply find stray calves or lambs in even the darkest or treacherous of conditions with TI?

In addition, this technology could greatly impact the way how animals are being treated. Detecting and treating certain ailments and injuries become easier with the technique of TI, which could possibly save on overall vet bills, reduce the usage of antibiotic, thus eventually save lives.

It is well known that keeping an eye on cows around a field is really a challenging task. Although, there is no need for real-time data. Recording the weight and health of cows when they come for milking prices would be helpful as it wastes no extra time and further assists in getting enough data to find trend lines and ensure productivity.

TI is also believed to provides a far safer method for keeping an eye on animals. If one ever needs to offer help to an injured animal, they will get to know just how prone they are to biting, kicking, and doing most anything they can in a way to safeguard themselves.

With the help of TI, one would be able to recognize the problem from a safe distance without even the requirement for a large supporting team. This could play an extremely imperative role in classifying the early stages of inflammation, mastitis, and lameness.

TI usually works by forming real-time thermograms formed by infrared radiation from an object (or animal, in this case), and so far, experiments are exhibiting a high level of accuracy. "TI cameras have been widely used to positively diagnose respiratory disease in calves a week before clinical symptoms arose!" reports the UK Daily Post.

While it is important to note that TI is still in its initial stage of development for animal use, it is already proving a successful technology for the farming industry.

The relatively new World Animal Health Information Database managed by the World Organization for Animal Health (OIE) is a significant database that tracks disease prevalence in alll regions of the World. In cooperation with the Food and Agriculture Organization of the United Nations (FAO), the OIE is investing disease rumors that surface on ProMED or other non-scientific sources of information; these early warning systems serve as a good alert system for emerging disease outbreaks.

The use of satellite-based remote sensing technologies could be useful as early warning systems for the emergence of serious infectious diseases, particularly those that are transmitted by arthropods. The FAO's Emergency Prevention System (EMPRES) for Transboundary Animal and Plant pests and Diseases program currently uses remote sensing technologies to determine the Normalized Difference Vegetation Index(NDVI) (Larkins et al., 2009).

Overall, whether "Agriculture 4.0" will be able to successfully accomplish its goal or not, or how quickly it arrives, it's a thrilling time for those who are associating with this industry. With technologies such as EID and RFID Tags, TI, and automated weighing technology, farmers are acquiring much-required visibility, feedback, and information that can only mean improved things for the health of their livestock and their bottom line.

2.3. BIOTECHNOLOGY FOR THE LIVESTOCK INDUSTRY

Meat and milk from farmed animals embracing livestock (goat, cattle, and buffalo) and poultry are sources of high-quality protein and vital amino acids, fats, minerals, and fatty acids, readily available vitamins, a trivial amount of carbohydrates and other bioactive components.

According to the Food and Agriculture Organization (FAO) 2008, it was estimated that there was a significant increase in meat consumption with an increase in population. The average global per capita consumption of meat stood at 42.1 kg/per year with 82.9 kg per year in developed and 31.1 kg per year in developing countries in a suggested daily animal-sourced protein per capita of 50 kg per year.

While, on the other side, milk is consumed in several forms: liquid, powder, cheese, butter, and cream at a global per capita consumption of 108 kg per person per year, which is far lower than FAO suggested daily consumption of 200 kg.

Some poor countries may not be able to cope with these levels of milk and meat requirement, resulting in malnutrition.

Demand for milk and meat production is also projected to twice in the year 2050 in developing countries, where the population will also likely to twofold. Thus, growing production and the effective marketing and secure processing of milk and meat, and their products are greater challenges for livestock producers.

Biotechnology is being exploited in several segments of the livestock industry to accelerate breed development for enriched animal welfare and health, improved nutritional safety and quality of animal-derived foods, and enhanced reproduction.

2.3.1. Reproductive Animal Biotechnology

It is generally seen that several biotechnology methods are used to enhance the breeding stock of animals. These comprise embryo transfer (ET), artificial insemination (AI), somatic cell nuclear transfer, *in-vitro* fertilization (IVF), and the developing technology on somatic cell nuclear transfer:

1. **Artificial Insemination:** One of the traditional perfected technologies is AI (AI) where new animal breeds are produced through the deposition of the male sperm from one high-class male to the female reproductive tract without the process of mating.

AI prevents the transmission of venereal disease, reducing the requirement of farms to preserve breeding males, assisting more precise recording of pedigrees, and reduces the cost incurred on introducing improved genetics. Several technologies have progressed that resulted in efficient use of AI in developing preferred livestock, including the approaches of cryopreservation or frozen semen and sperm sexing.

2. ***In-vitro* Fertilization:** In case when other artificial reproductive techniques do not give the desired results because of some difficulties such as non-responsive ovaries in the females, blocked reproductive systems, marginal semen quantity and quality in the male, and the existence of the disease, *in vitro* fertilization (IVF) is used. The fertilization of the egg and the sperm is conducted *in vitro* (outside the body of an animal) at definite biochemical and environmental conditions.

Till the present date, successful IVFs have been done in several animal species because of advancement in cryopreservation of reproductive cells and embryo production. Since the birth of the first rabbit through the technology of IVF in 1959, IVF offspring have been born in rats, mice, hamsters, guinea pig, cats, squirrels, cows, pigs, monkeys, and humans.

3. **Embryo Transfer (ET):** It is a transfer of embryo from one mother to a surrogate mother opens the pathways for the production of various livestock progenies from a high-class female. Selected females are persuaded to superovulate hormonally and inseminated at an adequate time in comparison to ovulation relied upon on the breed and species.

It is generally seen that week-old embryos are flushed out of the donor's uterus, quarantined, inspected with the help of microscopically for quality and number, and implanted into the lining of the uterus of surrogate mothers.

ET enhances the reproductive rate of chosen females, decreases the rate of disease transfer, and assists in the development of rare and economically imperative genetic stocks as well as the production of numerous closely allied and genetically comparable individuals which are significant in livestock breeding research.

According to an International Embryo Transfer Society (IETS), it was estimated that a total of almost 550,000 *in vivo* derived bovine embryos, 1,000 goat embryos, and 68,000 sheep embryos were transported worldwide in the year 2004.

4. **Somatic Cell Nuclear Transfer (NF):** It refers to a technique in which the nucleus (DNA) of a somatic cell is transmitted into a female egg cell or oocyte in which the nucleus has been detached to produce a new individual, which will be genetically similar to the somatic cell donor.

This technique was used to produce Dolly from a distinguished adult mammary epithelial cell which exhibited that genes that are already disabled in distinguished tissues can be wholly revived. Nuclear transfer technology provides the opportunity to generate clones from high-class genotype and can be made into use to effectively assess the effects of genotype x environment interactions and dissemination or testing of transgenics. Problems on high chances of pregnancy loss, newborn survival, and augmented frequency of abnormal development because of improper reprogramming of nuclear DNA (epigenetic inference) and rare conditions at the time of *in-vitro* processes make this a pre-commercial technology.

5. Embryo sexing & cloning:- It has been suggested that, if multiple sexed-embryo transfer became as routine an operation as AI is, beef operations based on this system could become competitive with pig and poultry production in terms of efficiency of food utilisation. Clones may be produced by embryo splitting and nuclear transfer. Cloning of embryos will not increase rates of genetic progress in the nucleus, but that it offers considerable advantages in increasing the rate of dissemination of tested superior genotypes in commercial populations. Other potential applications of cloning include efficient evaluation of genotype x environment interactions and testing and/or dissemination of transgenics (Patil and Patil, 2021).
6. Hormone use:- Use of hormonal assays to monitor reproductive function can be rewarding for both research purposes and commercial livestock operations. Reproduction can also be manipulated using hormonal treatments. Hormones produce desirable effects in animals. Lack of Knowledge about use of hormones & less economic viability of hormones limits the utility. Use of hormones leads to increase ovulation rate, induction of estrus, improved fertilization rate etc.
7. Multiple ovulation embryo transfer (MOET) and open nucleus breeding system:

Multiple ovulation embryo transfer (MOET) is a composite technology which includes superovulation, fertilisation, and embryo recovery, short-term in vitro culture of embryos, embryo freezing and embryo transfer.

Benefits of MOET

1. Increasing the number of offspring produced by valuable females.
2. Increasing the population base of rare or endangered breeds or species.
3. ex situ preservation of endangered populations.
4. Progeny testing of females.
5. Increasing rates of genetic improvement in breeding programmes.

Open nucleus breeding system

The ONBS concept is based on a scheme with a nucleus herd/flock established under controlled conditions to facilitate selection. The nucleus is established from the "best" animals obtained by screening the base (farmers) population for outstanding females. These are then recorded individually and the best individuals chosen to form the elite herd/flock of the nucleus. If ET is possible, the elite female herd is used through MOET with superior sires to produce embryos which are carried by recipient females from the base population. The resulting offspring are reared and recorded and the males among them are evaluated using, as appropriate, the performance of their sibs and paternal half sibs and their own performance. From these, an elite group of males with high breeding values for the specific trait is selected and used in the base population for genetic improvement through natural service or AI (Patil and Patil, 2021).

2.3.2. Genomics and Marker-Assisted Selection (MAS) Applications

The discovery and identification of molecular markers or DNA sequences connected with imperative animal traits have several purposes that comprise of heritability determination, trait improvement, and product traceability:

1. **Molecular Marker-Assisted Introgression (MAI):** Markers are made into use to guide livestock breeders in choosing individuals articulating the introgressed gene. A series of backcrossing to the recipient parent is typically prepared in traditional breeding style.

   With the use of molecular markers, the number and time of backcrossing cycles encountered in the identification and selection of the desired individual are abridged. In the present time frame, molecular markers are generally made into use in several livestock trait improvement activities such as wool quality, meat quality, growth, milk production, and quality, and disease resistance.
2. **Parentage, Product Traceability, and Genotype Verification:** It is generally seen that molecular markers are consistent tools used by regulatory bodies to keep an eye on food safety and product quality.

   Livestock parentage and its products can be traced and recognized with the help of molecular markers from farm to the abattoir, and from the cut-up carcass to the consumer's plate. An identical type of DNA-based technology has also been developed to identify the existence of around 211 bp fragments to enable testing of extremely minor meat samples from the supermarket.
3. **Screening for Undesirable Genes:** Genetic diseases, as well

as physical defects, can be documented and traced in livestock animals with the help of molecular markers. The origin and cause of these challenges can be easily traced to the DNA mutations and genetic changes as they are apparent in the protein function and structure. With DNA testing, animals transmitting these defective genes are certainly recognized and are culled from the livestock breeding program.

2.3.3. The Future of DNA-Based Technology in Livestock Improvement

In the current scenario, whole genomic sequences of imperative farm animals such as that of bovine and chicken have been freed, and genomic sequences of goat, pig, and sheep are now in the development phase.

With the advancement in sequencing farm animal genome, the ongoing development in molecular marker technology, and the use of reproductive biotechnology, prospects of research opportunities will be unfolded to enhance and modernize the livestock industry.

In the coming few years, it is likely to attain information and data on the genetic constitution of the animals that will permit a projection of the production ability of an animal at birth, or maybe even as a fetus, as well as the identification of animals best suitable to an explicit production environment.

2.4. TECHNOLOGICAL INNOVATIONS IN ANIMAL PRODUCTION RELATED TO ENVIRONMENTAL SUSTAINABILITY

Within the past few years, there has been an increase in concern across the world about the environmental damage prompted by intensive animal feeding operations. This augmented public concern has initiated some countries to introduce legislation (Figure 2.3) (JONGBLOED and LENIS, 1998).

The main area of distresses in context to large scale production of animal is the release of ammonia (NH_3) and Greenhouse gas emissions (GHG) that are associated with air quality degradation and contribute to global warming in addition to the impact of nutrient losses on water quality. For instance, the global dairy sector adds 4.0% of globally formed anthropogenic GHG, with a high proportion of such gases produced on the dairy farm (FAO, 2011).

Figure 2.3. Record keeping in animal husbandry.

Source: *Image by Broadview University.*

Methane (CH_4) is one critical GHG gas released from dairy farms, which is almost 25 times more effective in catching heat in the atmosphere compared to carbon dioxide (CO_2) over 100-years (Intergovernmental Panel on Climate Change (IPCC), 2006).

The major source of nutrient pollution from animal production is P and N excretion over crop removal. Although N is a dominant element for animals and plants, it is highly concentrated and highly mobile can be toxic and cause degradation in the environment. Phosphorus is a critical nutrient that plays a vital role in agricultural development.

Although, its release in the environment contributes to the depletion of non-renewable inorganic P sources and causing eutrophication. According to Steen (1998), it was projected that phosphate reserves would be depleted in the next 50 to 100 years.

Some of the technological innovations that can be taken into usage to mitigate environmental pollution and increase the sustainability of animal production are discussed as follows.

2.4.1. Greenhouse Gas Emissions

It is generally seen that GHGs absorb solar radiation that falls on the surface of the earth, contributing to the global warming effect. GHG emissions conducive to the agricultural sector that are associated with animal production comprises CH_4 right emitted from domestic animals, N_2O and CH_4 emitted from grazed lands and manure, and N_2O released from soils after application of manure (Kebreab et al., 2006).

Livestock production, particularly ruminants, contributes to CH_4 emissions globally at a considerable rate. Methane is also allied with dietary energetic losses that usually fall between 2% and 12% of the ingested gross energy (Johnson and Johnson, 1995).

Carbohydrates are agitated in the rumen, generating comparatively large amounts of CH_4 in comparison to the non-ruminant animals. Agriculture accounts for 50% of anthropogenic CH_4 emissions in the world (IPCC, 2006) and beef cattle were the prime contributor of enteric CH_4 emissions in the U.S., stood at 71%. Dairy cattle accounted for 25%, and the remaining were the result of the enteric fermentation of other animals (Environmental Protection Agency (EPA), 2007).

2.4.2. Enteric Fermentation

Methane in ruminants is often released during the process of fermentation of substrates in the rumen, particularly structural carbohydrates, by the exploit of a group of Archaea denominated methanogens. Methanogens are composed of *Methanobacterium*, *Methanobrevibacter*, *Methanosarcina*, and *Methanomicrobium* (Balch et al., 1979).

The majority of the methanogens make use of hydrogen generated in the formation of volatile fatty acids (VFA) in the rumen fermentation process, to lessen CO_2 and CH_4. However, some methanogens can make use of other substrates to yield CH_4, such as the employment of formate as a hydrogen source (Hungate et al., 1970).

Protists (often known as protozoa) also have a critical role in the CH_4 ruminal production since methanogens can be attached to protozoa and transmit hydrogen (Sharp et al., 1998).

Mitigation strategies intended at lessening CH_4 emissions from ruminants have become a critical tool to diminish the environmental impacts of livestock production and also to enhance the energetic efficacy of feed utilization.

Martin et al. (2010); and Hook et al. (2010) evaluated mitigation strategies aimed at management and biotechnological techniques, and dietary manipulation. According to Martin et al. (2010), best CH_4 mitigation strategies must diminish emissions by altering hydrogen production and exploitation without reducing nutrient digestibility.

The utilization of hydrogen should be shifted towards the creation of other biochemical compounds to keep a balance of rumen pH and not lessen the carbohydrate fermentation rate. In addition, mitigation strategies should be cost-effective to be adopted by the producers on a large scale.

Methane abatement technologies can be categorized into four chief types: use of ionophores and other compounds such as organic acids, dietary manipulation, use of biotechnologies, and defaunation.

2.4.3. Dietary Manipulation

Diet composition plays a very significant role in CH_4 production in the rumen. The synopsis of VFA formed in the rumen is reliant on the form of dietary carbohydrates fermented. Ruminal fermentation of structural carbohydrates endorses a better production of acetate, in which hydrogen is produced and can be taken into use by methanogens to produce CH_4.

The fermentation of non-structural carbohydrates errands the propionate production, which serves as a proton sink to sequester hydrogen. It is generally seen that forages generally have a higher composition of structural carbohydrates, so, are expected to enhance the share of acetate and obtainability of hydrogen for CH_4 production.

Although, dietary manipulation only has a restricted role in decreasing CH_4 emissions, particularly for lactating dairy cows, because a minimal quantity of fibre in the diet is needed for stimulation of saliva production, preservation of chewing activity, and buffering the rumen pH.

The physical form of Structural carbohydrate is also imperative because fibre from grains and by-products are generally not as efficient as fibre from forages in inspiring chewing activity and maintaining the overall fat percentage of milk (Armentano and Pereira, 1997) by maintaining levels of acetate in the rumen.

Excessive levels of dietary non-structural carbohydrates can result in higher VFA production that may surpass the absorption capability of the rumen wall.

The increased VFA production lessens rumen pH and encourages the production of lactate that further lessens rumen pH, ruining rumen bacterial activity and raising the occurrence of ruminal acidosis (Plaizier et al., 2008).

Since high-quality forage, such as young plants, produces more easily fermentable carbohydrates and less NDF, it has a higher digestibility and passage rate, reducing CH4 output by changing the fermentation pathway. More mature forage, on the other hand, produces more CH4 due to a higher C: N ratio, which reduces digestibility (Haque, 2018).

Gaviria-Uribe *et al.* (2020) also found that therapies with higher nutritional efficiency, such as higher digestibility, higher CP content, and lower FDN and FDA content, had higher DMI and lower CH4 emissions, and therefore lower CH4 energy loss. They also found that the intensity of CH4 emissions produced by the legume-based system was lower, making these systems a viable choice for transitioning to sustainable tropical cattle production.

According to research conducted by Dong *et al.* (2019); Jiao *et al.* (2016) and Mizeck *et al.* (2010), increasing the proportion of concentrate in cow ration decreases the amount of methane generated per kg of milk produced. In general, this occurs since a high-concentrate diet contains more readily fermentable substances (e.g., starch) than a high-forage diet.

Fischer *et al.* (2020) reported that precision feed restriction reduced daily methane emissions more for the least efficient cows by improvement of efficiency without impairing cow's performance.

2.4.4. Ionophores and Other Compounds

The use of ionophores in ruminant diets has been reviewed in the context of the upgrading of feed utilization (Russell and Strobel, 1989) and handling of ruminal fermentation, reducing the chances of acidosis (Tung and Kung, 1993). Although, with augmented worry about GHG emissions, the potential to lessen CH_4 emissions in the rumen by ionophores, particularly monensin, is of greater interest.

Monensin changes the transport of ions, particularly sodium, with the help of cell wall membrane, instigating an inhibitory effect in gram-positive bacterial population. The lessening of CH_4 emissions with the monensin use has been ascribed to the inhibitory effect on ruminal methanogens and in the lessening of hydrogen supply for the population of methanogenic.

Hook et al. (2009) reviewed the long-term impact of monensin and after 6 months, the diversity and quantity of ruminal methanogens were not impacted by monensin. Consequently, the plunge in CH_4 emissions with monensin supplementation appears to be more an outcome of the inhibition of gram-positive bacteria in the rumen, lessening the source of hydrogen to the methanogenic population, and also changing VFA from acetate to propionate.

Odongo et al. (2007) reported an overall 7% decline in CH_4 emissions without compromising milk yield or intake after six months supplementing monensin at 24 mg per kg of diet DM.

Rawat (2021) reported that the plant extracts/phytogenics reduce the production of methane via one of several ways such as inhibiting methane-producing bacteria, the protozoans that live in synergy with them, or altering/promoting the metabolic pathways involved in using the hydrogen that would otherwise be turned into methane.

Anggraeny *et al.* (2021) discovered that using a mixture of organic components and probiotics resulted in the greatest reduction in CH4 development from enteric fermentation.

2.4.5. Defaunation

It is important to note that the complete or fractional elimination of protists from the rumen is known as defaunation. The defaunation role in the reduction of CH_4 emissions is generally relied upon the hydrogen supply to the methanogenic population by rumen protists by using the hydrogen transmission process (Sharp et al., 1998).

Defaunation of ciliate protists in the rumen of steers enhances the overall concentration of propionate, whereas, on the other hand, steers with a usual protist population had a bigger concentration of butyrate in the rumen (Whitelaw et al., 1984). Methane emissions were abridged from 11.5 to 6.68 MJ/100 MJ of GE with defaunation.

Several substances and additives can endorse rumen defaunation, as discussed above (lipid supplementation, monensin) and results are generally relied on diet composition (Hegarty, 1999).

The rumen defaunation feasibility can be compromised with the alteration potential of rumen protists (as discussed for monensin). Furthermore, the defaunated animal's maintenance can be a challenging task (Bird et al., 2010) and nutrient digestibility can be weakened with the drop of rumen protists.

2.4.6. Biotechnologies

Various biotechnologies have been tested to lessen CH_4 production level in the rumen. Wright et al. (2004) abridged CH_4 production level by almost

8% by making use of a vaccine to quell methanogens. The progress of these vaccines dependent upon the cultivation and identification of different strains of methanogens, although, not all methanogens have been cultivated or recognized (Whitford et al., 2001).

Variances in rumen methanogens because of feed intake, climate conditions, and other facets in a different area of the world could restrict the vaccination activities in various regions in which the vaccines were developed (Wright et al., 2007).

Cooprider et al. (2011) attained a 31% lessening in CH_4 emissions from feedlot cattle by placing them with 14 mg estradiol benzoate, and 100 mg trenbolone acetate and also complementing monensin to the diet.

Other technologies, such as bacteriocins, passive immunization, were discussed by Martin et al. (2010) and often, outcomes are not determined over time or consistent between experiments.

2.4.7. Manure Management and Land Application

Alternative uses of manure can be detached into three types: adaptation to value-added products; conversion and use as an energy source; and emerging and innovative products. Many strategies and technologies are already in place for altering animal manure into energy and other esteemed products.

Newer technologies are also being developed with new products and improved efficiencies. Energy products that can be extracted from manure comprise of biofuels and heat, which includes liquid (e.g., alcohols from fermentation), gaseous (e.g., biogas from anaerobic digestion), and solid (e.g., densified pellets produced from fibres) fuels. Biological and thermochemical conversion technologies can be exploited to generate bio-energy while handling livestock wastes.

2.4.8. Anaerobic Digestion

Animal manure has intricated chemical composition and greater moisture proportion, thus, biogas production from anaerobic digestion seems to be the most energy-efficient preference (CAES, 2008). An appropriately functioning anaerobic digester can deliver several advantages at local, farm, and environmental levels.

These advantages include reduction of nuisance gas emissions; odour control; lessening of wastewater potency (oxygen demand); potential pathogen kill; conservation of plant nutrients (e.g., N, P, K) for use as a superior quality fertilizer; transformation of organic N into plant-available NH_3-N; and manufacture of a renewable energy source-biogas, that generally comprises of 50–70% CH_4 with the rest being chiefly CO_2 and impurities. Biogas can be advanced to biomethane by eradicating moisture, CO_2, and other impurities and conveyed to current and future natural gas distribution systems (Cantrell et al., 2008). The difficulties comprise of and high standards of management and maintenance and high initial capital investment.

2.4.9. Covered Lagoon

One of the most widely adopted ways to store manure from large animal facilities is using open-air, anaerobic lagoons. Though these requires low maintenance systems, they are conducive to emissions for GHG such as $CO_2$and CH_4, and other odorous intermediate compounds such as $NH_3$and hydrogen sulfides. Covering the lagoon with the help of a floating impermeable cover can lessen air pollutants and GHG, and harvest biogas (Desutter and Ham, 2005). Several governments of developing countries recognized the need of covering the lagoon as GHG reducing technology and farmers can also get carbon credit for covering their manure lagoons and spreading the biogas imprisoned under the cover (Climate Change Central, 2011).

2.4.10. Algal CO_2 Removal

It is important to note that CO_2 is a chief component in the product gases from thermochemical transformation processes and anaerobic digestion. Since an augmented atmospheric concentration of CO_2 is conceived as one of the key causes of global warming (Schneider, 1989), it is imperative to recover CO_2 to restrict short-term release. By innately fixing atmospheric CO_2 through the process of photosynthesis is 10 times more effective in comparison with the terrestrial plants (Usui and Ikenouchi, 1997), algae can promptly produce both intracellular oil and algal biomass (Miao and Wu, 2006). These algal products can then be converted and garnered into multiple value-added products.

2.5. APPLICATION OF INFORMATION TECHNOLOGY IN ANIMAL HEALTH

It is well known that livestock wealth is very crucial for various countries, especially for developing ones. Nowadays, even in developing countries, animal husbandry is no longer a subsidiary of agriculture or a backyard vocation (Figure 2.4).

Figure 2.4. Scanning to detect eating habits of cows.

Source: *Image by Plug and Play Tech Center.*

Animal husbandry has transformed into an independent industry and the modern reports propose that the animal husbandry sector contribution to the GDP is significantly higher irrespective of the meager input. Animal husbandry provides a better opportunity for marginal farmers whose income from agriculture is declining steadily because of the vagaries of monsoon, pest problems, fragmentation of landholdings, poor pricing, etc.

Though the progress of the livestock industry is very favourable, there is a need to integrate it with developments in other fields to get the maximum benefits. It is well known that the advancement in Information Technology over the past several years is incredible and provides huge potential in improving animal health through several measures such as efficient disease forecasting, modern therapeutic measures, and accurate and rapid and disease diagnosis, etc.

Veterinary informatics refers to the discipline concerned with the application of information science, computer technology, and engineering to support veterinary health care. Veterinary informatics and the larger field of medical informatics is generally called biomedical informatics or health care informatics and forms part of the broader domain of e-health. Applications of Information Technology in several aspects of animal health are discussed below.

2.5.1. Geo-Informatics Technologies in Animal Disease Surveillance

It is generally seen that initial identification of an infectious disease occurrence is a critical first step towards applying effective disease interferences and lessening resulting morbidity and mortality. Both seasonal and geographical distributions of several infectious diseases are associated with climatic conditions, thus the likelihood of making use of seasonal climate forecasts as projecting indicators in disease early warning systems (EWS) has long been a major area of interest.

Remote sensing (RS), geographic information system (GIS), and global positioning system (GPS) are the three widely adopted veterinary geo-informatics technologies engaged in this information era for quick worldwide communication of data for the organization of animal diseases.

GIS can be defined as a computerized database management system for storing, capturing, checking, manipulating, integrating, analyzing, and displaying data in context to location.

GPS is a data collection technology, whereas on the other hand, GIS is a data analysis technology. What divides GIS from other types of databases/information is that everything is dependent on location (georeference). This georeferenced information is nothing more than "where things are," such as longitude and latitude coordinates.

Today, it is usually seen that GIS is being employed for disease monitoring in the United States of America. The Veterinary Services of the United States Department of Agriculture (USDA exploits the advantages of integrated and comprehensive surveillance and monitoring to validate where a specific disease does or does not arise.

For instance, one GIS map can demonstrate the population density of any agricultural species with the help of National Agricultural Statistics Service census data. A second map can demonstrate entire disease monitoring testing done in a particular timeframe.

An assessment can then be done to make sure that disease testing and monitoring resembles the population density of that species in the area. Lastly, a third map can be made to represent changes or areas to which the focus should entail or future monitoring purpose. It will help by convincing that funds that are used for monitoring purpose are being spent in the right areas. Along with GIS, RS has transformed the way scientists manage and evaluate geographic data. RS can be defined as the acquisition of geographic data without initiating physical contact with the area of study. In the current time frame, RS normally relates to aerial imaging/photography and satellite images, both of which are subtle to the proportion of the electromagnetic spectrum not perceptible to the human eye or even majority of cameras (such as microwave, infrared). The potential to gauge concealed energy and to record huge sums of high-resolution data from unlimited distances is an extremely dominant tool.

2.5.2. Information Technology in Disease Diagnosis

It is important to note that medical diagnostic technology has made rapid strides with the advancement in computer technology. Much of the progress being made in human diagnostic technologies is decoded into veterinary medicine in developed countries.

Newer branches such as radiodiagnosis, imaging, telemedicine, teleradiology, and telesonography have emerged. Generally, the devices/telesonography which has been produced with contemporary technologies in the current digital age are discussed as follows:

1. **Image Intensifier TV System (IITV):** This system is mostly used in orthopaedic surgery. This is basically concerned with repairing fracture with the help of a small incision thus attaining minimal invasive surgical manoeuver. IITV generally assists in X-ray imaging of the intraoperative site for orthopaedic handlings, and the alike can be saved for future reference.
2. **Ultrasound:** In small animal as well as equine practice, ultrasound is regularly used as a diagnostic aid. Applications of ultrasound in ruminants have not been exploited fully, except in pregnancy. It is well known that there could be several organs that could be scanned by an ultrasound scanner.

It seems like Ultrasonography has a promising future in veterinary medicine, especially for the assessment of intra/peri-abdominal disease. Ultrasonography is considered as the solo most versatile add-on to the non-surgical and non-invasive armamentarium of the veterinary clinician since the introduction of the fiber optic endoscope.

3. **Computerized Tomography (CT):** It has been tremendously substantial progress which has a unique cross-sectional imaging ability potential for diagnosis of malformations, tumours, degenerative, inflammation, and vascular trauma and diseases.

CT is a diagnostic modality that is distinctive from a simple X-ray; an organ is scanned in succeeding layers by a narrow beam of X-rays in such a manner that the transmission of X-ray across a specific layer can be computed by a computer, and used to frame a picture of the internal structure.

4. **Magnetic Resonance Imaging (MRI):** It is a highly sensitive and non-invasive technique that plays a very important role in providing comprehensive and accurate anatomic images with spatial resolution and good contrast.

Although, in veterinary medicine, MRI is still considered as infancy and its usage is very rare. Till the present day, MRI has been widely used in developed nations in clinical cases as well as a research tool particularly for central nervous system (CNS) diseases in small animals.

It is important to note that MRI has a broader range of applications. It is used for imaging entire body parts in small animals, but only the edges and in the large animals; the only head can be an image.

The newer applications of MRI are MR spectroscopy and MR angiography. It is particularly used to distinguish tumours from peri-tumoral oedema and an inflammatory process from a neoplastic mass. It is more sensitive and specific in localizing, detecting, and distinguishing cellulitis, osteomyelitis, and abscess.

5. **Nuclear Scintigraphy:** It is considered a highly sensitive innovative procedure in which radioisotopes are taken into use to identify the functional abnormalities of the body system. The interpretation is relied upon the presence of the decreased (cold spots) or increased (hot spots) radioactivity regions.

For instance, an active process is specified by a hot spot while a dull process such as lack of perfusion is signified by a cold spot. Nuclear scintigraphy has been taken into use to look for functional disorders of the kidney, lungs, liver, thyroid gland, GI tract, and many other organs.

It is extremely important in the diagnosis of lung perfusion, occult lameness, and patency and ventilation of the ureter in both small as well as large animals. In addition, used for vertebral column monitoring and imaging the development of fracture healing and in the detection of tumour.

6. **Digital Subtraction Angiography (DSA):** It is a radiographic modality that provides the opportunity of dynamic imaging of the vascular system succeeding intravascular injection of iodinated X-ray contrast media, with the help of using image intensification,

augmentation of the digital processing and iodine signal of the image data.

Temporal subtraction of the images attained at the time of first arterial phase of injection of the contrast medium from the images attained before and after juxtapose medium administration harvest images that are barren of soft tissue and iodine signal. This imaging modality is very important as it plays a crucial role in emphasizing vascular pathologies such as stenosis, etc.

7. **Laparoscopy:** It is well known that laparoscopy has been an appreciated therapeutic and diagnostic tool in human clinical medicine. Only in the past decade, its use has been considered in several animal species for research and clinical therapeutic and diagnostic purposes. The most beneficial quality of laparoscopy is that it permits express inspection of the abdominal cavity with only negligible and artificial surgical intervention.

Thoracoscopy has been engaged in humans for the treatment and diagnosis of diseases of the lung, pleura, great vessels, mediastinum, oesophagus, and pericardium.

Visceral inspection of the thoracic cavity by thoracoscopy has been used to ensure a more precise prognosis and diagnosis in horses affected with pleuropneumonia and other oesophageal and thoracic disorders. Thoracoscopy open the doors of biopsy and visualization of a large surface of the lung and offers pertinent specimen for histopathological diagnosis.

8. **Endoscopy:** It is a nominal invasive diagnostic modality that helps in documenting mucosal inflammation, active bleeding, hyperemia, an uneven mucosal surface, etc., and assists biopsy in tubular organs such as respiratory, GI tract, and the urogenital systems.

9. **Pulse Oximetry:** It carries the explicit importance of constantly checking the saturation of haemoglobin with oxygen effortlessly and noninvasively, thus offering a measure of cardio-respiratory purpose.

The central physical property that permits the pulse oximeter to compute the oxygen saturation of haemoglobin is that blood alters its color as haemoglobin sponge up erratic amounts of light relied on its saturation with oxygen. Henceforth, pulse oximetry remains the standard of care during the process of anaesthesia as well as in the intensive care unit and recovery room.

2.5.3. Artificial Intelligence in Health Management

Artificial intelligence can be defined by human functions and comparing computer. If the computer has done a task that seems like it is almost done by a person, for instance, in an intelligent manner, it can be said to be exhibiting artificial intelligence.

In the field of medicine, the majority of artificial intelligence research has been dedicated to forming computer systems that embrace comprehensive information about a particular medical subject.

By concentrating on pertinent information on the problems confronting the physician, these programs are intended to act as counsellors and thereby possess the ability to augment the practitioner's expertise and experience in the concerned field.

Expert systems are generally referring to computer programs that usually comprises of a large chunk of knowledge for making decisions about precise problem domains such as an area of medicine. In medicine, several imperative experimental expert systems have been established. For instance:

- PIP-Renal disease;
- Internist-Diagnosis in internal medicine;
- VM-Ventilator Management;
- ATTENDING-Anesthetic Management;
- PUFF-Pulmonary function.

2.5.4. Information Technology in Instrumentation

Another significant region of the contribution of Information Technology is in the instrumentation side. The contemporary molecular methods of diagnosis needed complex electronic equipment. It is generally unreasonable for a researcher to possess all the needed information and data in the field of electronics to handle this equipment.

To shorten this, all the modern equipment such as high-performance liquid chromatography (HPLC), ELISA reader, UV Spectrophotometer, RIA, Atomic Flowcytometer, Absorption Spectrophotometer, ultra-low freezer, freeze drier, polymerase chain reaction (PCR) machine, etc., are now organized by microcomputers and user-friendly software are in place to handle them. With this software, even a novice person can start managing this equipment with nominal training.

2.5.5. Information Technology in Data Analysis

It is generally seen that analysis of data is another significant area in animal health. The data which has been amassed from outbreaks over a certain period of time has to be statistically examined to form models and project future incidences of the disease.

From what we know so far, manual analysis is error-free, although being time-consuming. Many simple and user-friendly statistical packages are in existence that can do a similar task in the same effective and efficient manner (Figure 2.5).

Figure 2.5. Putting tags on each cow for identification and data analysis.

Source: *Image by Pixabay.*

2.5.6. Information Technology in Molecular Biology

One of the most significant areas of animal health is the examination of genomes. Genomic analysis gains significance since it is the indisputable form of diagnosis. To advance gene probes, the whole genome has to be planned. After the full plan, there is a need to identify the unique areas in comparison to the other genome of other infectious agents.

Though the genome is insignificant in size, the number of bases in each genome is afar manual comparison. There is much software in existence that can compare the genome with prevailing genomes in a gene bank to classify rare areas. Such software also enables the opportunity for developing a molecular clock, phylogenetic tree, and for developing primers for PCR reaction.

The molecular clock and phylogenetic tree are necessary for molecular biologists to reach at conceivable origins of the virus. This finding carries more weight if the virus strains are immunologically distinct.

2.5.7. Information Technology in Simulation Studies

Another area where IT can deliver a huge contribution is the simulation of *in vivo* conditions. There is some advanced graphics-based software that would work best in this case. Transferring of etiological agents, proteins, etc., in between the cells can be reviewed with the help of this software. Study on these trafficking patterns is important to form vaccines for intracellular bacteria and cell-associated viruses.

The function of graphics-based software in examining the 3-D structures of antibody and antigen is also essential to study the relations between them *in vivo*. Interaction of antibody and antigen is necessary to study the pathogenesis of any infectious diseases. In addition to this, graphics-based software also provides unlimited scope as educational aids.

Apart from what is discussed above, Information Technology contributes considerably in several other areas and is bound to play an imperative role in improving the overall health of an animal; thereby profiting a mainstream of our population which is reliant on agriculture and similar activities.

2.6. CONCLUSION

At the end, it is concluded that with the advancement in technology, there has been significant development seen in animal husbandry. Animal husbandry is one of the most important sectors as it greatly contributes to the GDP of the economy, and it is essential for the overall welfare of society.

There are a lot of challenges that are associated with animal husbandry, such as improper health management, inadequate feed intake, prevalent uncertain disease, and ecological imbalance, uncertain climatic conditions.

These challenges are likely to increase in the future, and it is the only technology that can effectively address these challenges. As these challenges are becoming more complicated, it is imperative for technology also to progress in a similar proportion.

In addition, to ensure effectively and efficiently management of animal husbandry, there is a need for cooperation and coordination among various independent group and countries concerned with the de elopement of this industry. By integrating technology and ensuring cooperation among various members, one can see development in this sector.

REFERENCES

1. Anggraeny, Y. N., Pamungkas, D., Mariyono, Krishna, N. H., Antarim R., Putri, A. S., & Apriliza, M. N., (2021). Evaluation of the use of plant organic components and probiotics on ruminal characteristics and as a decrease of methane. *IOP Conference Series: Earth and Environmental Science,* 648 012191.
2. Armentano, L. &, Pereira, M., (1997). Measuring the Effectiveness of Fiber by Animal Response Trials, Journal of Dairy Science, 80(7):1416-1425.
3. Larkins, B.A., Brigs, S.P., Delmer, D.P., Dick, R.P., Flavell, R.B., Gressel, J., Habtemariam, T., Lal, R., Pell A.N., Leger, R.J.S., & Wall, R.J., (2009). National Research Council, 2009. Emerging Technologies to Benefit Farmers in Sub-Saharan Africa and South Asia. [online] Available at: <https://www.nap.edu/catalog/12455/emerging-technologies-to-benefit-farmers-in-sub-saharan-africa-and-south-asia#rights> [Accessed 11 May 2021].
4. No author, this is Government website (2019).*Animal Husbandry Including Management*. [eBook] Available at: http://apsc.nic.in/pre-animal-husbandry-including-management.pdf (accessed on 15 March 2021).
5. No author it is from government website(2021).*Animal Husbandry and Dairying*. [eBook] Available at:https://niti.gov.in/planningcommission.gov.in/docs/plans/planrel/fiveyr/10th/volume2/v2_ch5_2.pdf (accessed on 15 March 2021).
6. Biolyse, (2021).*New Technology Used in Raising Livestock*. [Online] Available at: http://www.biolyse.ca/technology-used-in-raising-livestock/ (accessed on 15 March 2021).
7. Bird, V., Premkumar, P., Kendall, T., Whittington, C., Mitchell, J., & Kuipers, E. (2010). Early intervention services, cognitive-behavioural therapy and family intervention in early psychosis: systematic review. *The British journal of psychiatry : the journal of mental science*, 197(5), 350–356. https://doi.org/10.1192/bjp.bp.109.074526.
8. Birsa Agricultural University, (2021).*Faculty of Veterinary Science and Animal Husbandry*-Birsa Agricultural University. [Online] Available at: https://www.bauranchi.org/faculty-of-veterinary-science-animal-husbandry/ (accessed on 15 March 2021).

9. Birthal, P. S., &Parthasarathy Rao, P, (2002).*Technology Options for Sustainable Livestock Production in India.* [eBook] Available at: http://www.icrisat.org/PDF/PPR-Livestock-report.pdf (accessed on 15 March 2021).
10. Cantrell, K.B., Ducey, T., Ro, K.S., Hunt, P.G., (2008). Livestock waste-to-bioenergygeneration opportunities. *Bioresource Technology.* 99:7941–7953.
11. Cooprider, K. L., Mitloehner, F. M., Famula,T. R., Kebreab, E., Zhao, Y., & Van Eenennaam, A. L., (2011). Feedlot efficiency implications on greenhouse gas emissions and sustainability, *Journal of Animal Science*, 89(8):2643–2656, https://doi.org/10.2527/jas.2010-3539.
12. DeSutter, T.M., Pierzynski, G.M. and Ham, J.M., (2005), Movement of Lagoon-Liquor Constituents below Four Animal-Waste Lagoons. *Journal of Environmental Quality*, 34: 1234-1242. https://doi.org/10.2134/jeq2004.0296.
13. Dong, L., Li, B., & Diao, Q., (2019). Effects of Dietary Forage Proportion on Feed Intake, Growth Performance, Nutrient Digestibility, and Enteric Methane Emissions of Holstein Heifers at Various Growth Stages. *Animals*. 9(10): 725. Doi: 10.3390/ani9100725.
14.
15. Elezaj, R., (2018).*Two Technologies Helping Agriculture and Animal Health.* [Online] Machine Design. Available at: https://www.machinedesign.com/automation-iiot/article/21837956/2-technologies-helping-agriculture-and-animal-health (accessed on 15 March 2021).
16. EPA. 2007. Inventory of U.S. Greenhouse Gas Emissions and Sinks: 1990–2005. Washington, DCU.S. Environmental Protection Agency.
17. FAO. 2011. Guidelines for the preparation of livestock sector reviews. Animal Production and Health Guidelines. No. 5. Rome.
18. Fay, T., & Ku, L., (2021).*Livestock Farming Technology in Animal Agriculture.* [Online] Plug and Play Tech Center. Available at: https://www.plugandplaytechcenter.com/resources/livestock-farming-technology-animal-agriculture/ (accessed on 15 March 2021).
19. Fischer, A., Edouard, N., & Faverdin, P., (2020). Precision feed restriction improves feed and milk efficiencies and reduces methane emissions of less efficient lactating Holstein cows without impairing their performance. *Journal of Dairy Science*. 103 (5):4408-4422.

20. Fmprc.gov.cn. (2021).*Science and Technology in Agriculture and Animal Husbandry*. [Online] Available at: https://www.fmprc.gov.cn/mfa_eng/ljzg_665465/3585_665495/3586_665497/3587_665499/t17932.shtml (accessed on 15 March 2021).
21. Gaviria-Uribe, X., Bolivar, D.M., Rosenstock, T.S., Molina-Botero, I.C., Chirinda, N., Barahona, R. & Arango, J., (2020). Nutritional Quality, Voluntary Intake and Enteric Methane Emissions of Diets Based on Novel Cayman Grass and Its Associations With Two *Leucaena* Shrub Legumes. *Frontiers Veterinary Science*. 7:579189. Doi: 10.3389/fvets.2020.579189.
22. Hamid, R., (2021).*Novel Technologies in Animal Husbandry*. [Online] www.technology times.pk. Available at: https://www.technologytimes.pk/2020/09/04/novel-technologies-in-animal-husbandry/ (accessed on 15 March 2021).
23. Haque, M. N., (2018). Dietary manipulation: a sustainable way to mitigate methane emissions from ruminants. *Journal of Animal Science and Technology*. 60:15. https://doi.org/10.1186/s40781-018-0175-7.
24. Hegarty R.S., (1999). Reducing rumen methane emissions through elimination of rumen protozoa. *Australian Journal of Agriculture Research*, 50:1321–1327.
25. Hook, S.E., Wright, A.D.G., & McBride, B.W., (2010). Methanogens: Methane Producers of the Rumen andMitigation Strategies. *Archaea,* Article ID 945785, https://doi.org/10.1155/2010/945785.
26. Hungate, R.E., Smith, W., Bauchop, T., Yu, I., & Rabinowitz, J.C., (1970). Formate as anintermediate in the bovine rumen fermentation. *Journal of Bacteriology,* 102:389–397.
27. Indiaeducation.net. (2017).*Animal Husbandry and Veterinary Science IFS Examination Syllabus*. [Online] Available at: https://www.indiaeducation.net/civil_services/upsc/ifs-exam/animal-husbandry.aspx (accessed on 15 March 2021).
28. IPCC (2006). 2006 IPCC Guidelines for National Greenhouse Gas Inventories, Prepared by the National Greenhouse Gas Inventories Programme, eds. HS Eggleston, L Buendia, K Miwa, T Ngara, K Tanabe. Japan: IGES.
29. Isaaa.org. (2021).*Biotechnology for the Livestock Industry* | ISAAA.org. [Online] Available at: https://www.isaaa.org/resources/publications/pocketk/40/default.asp (accessed on 15 March 2021).

30. Johnson, K. A., & Johnson, D. E. (1995). Methane emissionsfrom cattle. *Journal of Animal Science*, 73:2483–2492.

31. Jongbloed, A. W., & Lenis, N. P. (1998). Environmental concernsabout animal manure. *Journal of Animal Science*, 76:2641–24648.

32. Kebreab, E., Moraes, L., Strathe, A., & Fadel, J., (2012). Technological innovations in animal production related to environmental sustainability. *Brazilian Journal of Animal Health and Production, 13*(4), 923–937.

33. Kebreab, E., J. France, B. W., McBride, N. E., Odongo, A., Bannink, J. A., Mills, N., & Dijkstra, J., (2006). Evaluation of models to predict methane emissions from enteric fermentation in North American dairy cattle. Pages 299-313 in Nutrient Utilization in Farm Animals: Modelling Approach. E. Kebreab, J. Dijkstra, W. J. J. Gerrits, A. Bannink, and J. France, eds. CAB International, Wallingford, UK.

34. Lal, R., (2020). Integrating animal husbandry with crops and trees. *Frontiers in Sustainable Food Systems*, 4.

35. Luvas.edu.in. (2021).*Veterinary and Animal Husbandry Extension Education*. [Online] Available at: https://luvas.edu.in/veterinary-and-animal-husbandry-extension-education.html (accessed on 15 March 2021).

36. Martin, C. Morgavi, D.P. & Doreau, M., (2010). Methane mitigation in ruminants: from microbe to the farm scale. Animal, 4(3):351-365.

37. Megahvt.gov.in. (2021).*Policy and Programs: Animal Husbandry and Veterinary Department, Government of Meghalaya*. [Online] Available at: http://megahvt.gov.in/policy.html (accessed on 15 March 2021).

38. Miao, X., & Wu, Q., (2006). Biodiesel production from heterotrophic microalgal oil. *Bioresource technology*, 97(6), 841–846. https://doi.org/10.1016/j.biortech.2005.04.008.

39. Mizeck, G. G., Chagunda, J. Flockhart, F., & Roberts, D. J., (2010) The effect of forage quality on predicted enteric methane production from dairy cows. *International Journal of Agricultural Sustainability*, 8:4: 250-256, DOI: 10.3763/ijas.2010.0490.

40. Nifa.usda.gov. (2021).*Animal* Production. National Institute of Food and Agriculture. [Online] Available at: https://nifa.usda.gov/topic/animal-production (accessed on 15 March 2021).

41. Ntt-review.jp. (2020).*Application of Information and Communication Technology to Agriculture and Animal* Husbandry. NTT technical review. [Online] Available at: https://www.ntt-review.jp/archive/

ntttechnical.php?contents=ntr201606fa3.html (accessed on 15 March 2021).

42. Patil, P.V., & Patil, M.K., (2021). *Milk production management*. Jaya publishing house, Delhi, India.
43. Plaizier, J. C., Krause, D. O., Gozho, G. N., & McBride, B. W., (2008). Subacute ruminal acidosis in dairy cows: the physiological causes, incidence and consequences. *Veterinary journal (London, England: 1997)*, 176(1): 21–31. https://doi.org/10.1016/j.tvjl.2007.12.016.
44. Rawat, S. (2021). Feed Additives Put the Brakes on Cattle Methane Emissions. https://www.labiotech.eu/in-depth/feed-additives-cattle-methane-emissions.
45. Russell, J. B., & Strobel, H. J. (1989). Effect of ionophores on ruminal fermentation. *Applied and environmental microbiology*, 55(1), 1–6. https://doi.org/10.1128/AEM.55.1.1-6.1989.
46. Schneider, S. H., (1989). The Greenhouse Effect: Science and Policy. *Science*, 243(4892):771-781.
47. Sharp, R., Ziemer, C.J., Stern, M.D. & Stahl, D.A. (1998). Taxon-specific associations between protozoal and methanogen populations in the rumen and a model rumen system. *FEMS Microbiology Ecology*, 26: 71-78. https://doi.org/10.1111/j.1574-6941.1998.tb01563.x.
48. Sites.google.com. (2019).*Livestock Production Management-Application of Information Technology in Animal Health*. [Online] Available at: https://sites.google.com/site/viveklpm/information-technology-in-veterinary-science/application-of-information-technology-in-animal-health (accessed on 15 March 2021).
49. Steen, I. (1998). Phosphorus availability in the 21st century: Management of a non-renewable resource. Phosphorus and potassium, 217: 25-31.
50. Tribe, D., (1993). Animal husbandry, animal production and animal science in Asia.*Outlook on Agriculture, 22*(1), 7–11.
51. Tung, R.S., & Kung, L., (1993). In Vitro Effects of a Thiopeptide and Monensin on Ruminal Fermentation of Soluble Carbohydrates1. *Journal of Dairy Science*, 76(4):1083-1090. https://doi.org/10.3168/jds.S0022-0302(93)77437-8.
52. Usui, N., & Ikenouchi, M., (1997). The biological CO2 fixation and utilization project by RITE(1): highly-effective photobioreactor system. *Energy Convers Management*, 38(Suppl 1):S487–S492.

53. Wankhede, D., & Pathakala, M., (2021).*Use of IoT in Animal Husbandry*. [eBook] Available at: http://www.iosrjournals.org/iosr-jce/papers/Conf.17025-2017/Volume-2/7.%2040-44.pdf (accessed on 15 March 2021).

54. Whitelaw, F.G., Eadie, J., Bruce, L., & Shand, W. (1984). Methane formation in faunated and ciliate-free cattle and its relationship with rumen volatile fatty acid proportions. *The British journal of nutrition*, 52 (2): 261-75.

55. Whitford, M.F., Teather, R.M. & Forster, R.J. (2001). Phylogenetic analysis of methanogens from the bovine rumen. *BMC Microbiology,* 1: 5. https://doi.org/10.1186/1471-2180-1-5.

Chapter 3

Techniques Used in Animal Husbandry

CONTENTS

Techniques are very important for the effective and efficient management of animal husbandry. In the chapter techniques used in animal husbandry, the technique of electronic identification (EID) and its role is explained. The chapter also explains the importance of breeding and genetics. It also shed thrown some light on reproductive performances.

Methods of animal breeding are also explained in this chapter. In addition, the concept of controlled breeding experiments is also explained in this chapter. In addition, some technologies that can improve productivity in the livestock industry are also explained. It also discusses the technique of electromechanical engineering for improving the livestock industry. Some novel technologies in animal husbandry are also discussed. In the end, the use of IoT in animal husbandry is discussed.

3.1. INTRODUCTION

Even as the number of farms is decreasing, due to an increase in the world population, the demand for livestock products that are reliable and have quality is seeing an increase resulting in an increase in both animal production as well as the number of animals per farm. There has been an additional increase in the problems with respect to livestock production. Multidisciplinary studies in varied fields like technology can provide a solution to these problems.

The expected performance from animals that have a high generic value in large enterprises can be increased to the expected levels through the use of automation systems and technology. Work relating to animal health, their welfare and production daily in livestock farming is fairly simple and requires standard application with data monitoring on a routine basis.

At the same time, the workload and error burden both see a rise with an increase in the number of animals. To be successful, farmers must improve upon production by exploiting the technological changes and adapting their infrastructure rapidly. In an environment of intense competition for the convenience of the user, automation systems and mechanism present varied options.

Many milking systems are being replaced by automated recording (milk conductivity, milk yield, body weight measurements, and activity recording) even though most of the data is currently collated manually as a result of which both the quality and quantity of data is better. Since 1980 there has been a rapid increase in the farms using automation systems.

Enhanced automation can benefit by and large-all large and medium-sized farmers. Automation systems and technologies have numerous opportunities for facilities. For the optimization of farm management decisions, robots are increasingly being used by farmers on algorithms or production. A new automation system is being created by technological developments as a result of which livestock production can see more flexible and smarter possibilities in the related work.

The challenges thrown by a competitive market can be met through automation of animal husbandry and integrating the processes and systems on-farm, which facilitates the entire process. Electronic milking, recording, auto-drafting, heat-detection auto-weighing, barn optimization, feeding, livestock housing, genetic improvement, equipment design and health monitoring are some of the main areas where technology can be used effectively.

The decision of the dairyman with respect to plans about the future of the dairy is made more convenient and easier through the numerous opportunities provided by technology. Various aspects of animal husbandry like meat production, milk, and the herds' reproductive performance are affected by the key innovations and latest challenges. This chapter deals with these matters through the valuable information on these aspects.

At the same time have research results that are detailed, the same have been summarized with innovative approaches regarding beef cattle breeding, reproductive performance characteristics, and dairy cattle. Systematic reviews have been introduced in this chapter, and the results of innovative research in animal welfare, their health, and animal husbandry have been discussed.

The scientific viewpoints that currently exist to the definition and concept of innovations in animal husbandry are attempted to be reviewed in this chapter. Further, there is a discussion on the matter of systematic reviews' usage with respect to questions about economy, usage, intervention effects, negative, and positive aspects of innovations and technology.

There is a stress on the requirement of using validated indicators for assessing innovation scientifically as also the necessity of having interaction between varied disciplines.

The role of innovation and the requirements of technology for getting a good and better quality of animal products in animal production have been examined initially in this chapter. Further, by using detailed research results, there has been a review on the usage possibilities of the equipment and the innovative technologies currently available. Thereafter, the varied technologies that can be used to gather production knowledge and technologies usage that is more convenient have been examined. Eventually, the overall level of innovation in the field of animal production has been assessed, which is based on the results of worldwide research.

Further, a few implications and suggestions about the side effects that were not intended in its application and production shall be summarized in addition to the benefits derived from innovation.

According to the efficiencies required for their usage in animal agriculture, new biotechnologies are being developed as well as refined about gene transfer, cloning, *in-vitro* production and sexing of embryos.

In-vitro production of embryos on a commercial basis has been made possible through efficient *in-vitro* systems for capacitating spermatozoa and maturing of oocytes for developing and fertilizing the embryos.

The embryos' supplied nuclear material has been used to achieve the cloning of embryos through nuclear transfer in the case of sheep, pigs, cattle, and rabbits; in sheep as late as the 120-cell stage. To enhance the efficiency of this procedure, a lot of research is still required even though embryos have been re-cloned. To increase the number of clones to thousands or even millions, research is required to be done on the usage of cultured cells in the embryo.

Currently, commercial testing is being done on amplified deoxyribonucleic acid (DNA) hybridization, which is a more efficient method for sexing the embryos of most species with male specific antibodies in a non-damaging way. For rats, mice, chicken, rabbits, cattle, pig, fish, and sheep offspring or transgenic embryos have been produced.

A better understanding of the genes that need to be transferred along with methods that are more efficient with better control over the tissue of the specific gene that needs to be expressed and its time are required even though as of now for the genes can be targeted for expression in specific tissues.

Knowledge about the genes to be introduced is essential before transgenic animals of value can be made in large numbers. At present, there is not much understanding about the genes that influence animal growth, environmental adaptation, the efficiency of growth, the composition of meat, egg, or milk, and the resistance of animals towards disease. The genome of domestic animals needs to be mapped, and efforts are required in this field so that their identification can be done. Within the next decade, livestock breeding can be changed drastically through these technologies as well as other newer ones.

3.2. ELECTRONIC IDENTIFICATION

In the 1970s, the EID system started. All the same, the readable, visual markings placed on the animals are dealt with by the current laws (EU Directives 820:97: EC and EU Directives 92:102: EEC). Livestock producers have several animal ID technologies available to them. In all probability for identification of cattle, RFID or Radio Frequency Identification (RFID) shall be used.

An electronic number is there in these devices which is unique for each animal, and the animal shall be linked to the database through this number. For animal identification, the latest technology includes boluses with transponders inside the reticulum, injectable transponders, and electronic ear tags. Further for the identification of animals, numerous kinds of RFID tags like ear tags, boluses, and injectable glass tags are placed subcutaneously.

Data is sent using radiofrequency for these systems to work. Boluses are considered to be safe for the health of animals as the first two parts of stomach of the ruminants retain them. The retention rate can reach 100% in the lambs, and they can be administered to them at the 5th week after weaning. On the other hand, even though the location preferred in the case of each animal species varies, after the birth itself, the injectable transponders can be applied.

Cattle farmers can easily avail these technologies (rumen boluses, ear tags, and implants) in the market. For evaluation of the data by the base computer, a special chip system is there in all these devices.

For evaluating herd data, specific components are present in the devices so that data can be stored and evaluated. A reader is present in many electronic tags so that data can be received and stored for evaluation.

For another evaluation stage, the number is transferred to another storage system by some of the tags. For transferring the data on the system antenna are used. On the technological side, as per the carrier frequency band, the grouping of RFID tags is done into two categories, namely: high frequency or HF tags functioning at 13.56 MHz and the low frequency (LF) or the LF tags that function at a frequency of 125–134.2 kHz.

The body condition score can be automatically determined through electronic scales. The electronic weighing system is another technology that is of great use to the farmers as it is a powerful and easy electronic weighing system by which cattle weight can be measured accurately. This enables the farmers to monitor cattle performance continuously and easily.

The cattle squeeze or the waterer was established by these systems on the road. The computer receives the stored information for evaluation. This system is aided by auto-drafting whereby there is the automatic separation of cattle going through the race based on sex, age, weight, or any other criteria that measures performance.

3.3. BREEDING AND GENETICS

The latest technology is essential in dairy farms so that the expected performance may be derived from the breeding animals that have a high genetic value. Dairy farming can extend tremendous advantages to the animals, farmer, and the consumer if the programs for the management of dairy cattle herd are effective. For electronic updating that is expanded, the type evaluation and genetic information of bulls and herd members are specifically useful.

Genetics is a branch of biology which is concerned with the study of genes, genetic variation, and heredity in organisms. Nevertheless, heredity had been observed for millennia, Gregor Mendel, Moravian scientist and Augustinianfriar working in the 19th century in Brno, was the first one to study genetics in a scientific way.

All the same, detailed knowledge of the functions and their effective use is essential to be able to derive this system's full advantage. Expected results shall not be obtained from the assessment or record keeping of the heavy data flow that results from the immense amount of data received concerning numerous animal-related issues, individual animals, and herd management unless it is used in decisions.

The use of gene technology and breeding in animal husbandry has brought about a tremendous change in the various breeds. Within-breed selection, crossbreeding and breed substitution has met the demand for livestock products till the 1980s.

However, new techniques like artificial insemination (AI) and selection techniques that are more specific need to be used for meeting future demands. In the livestock sector, the possibility of a higher rate of genetic gain can be provided by genomic selection.

In the future phenotypic and pedigree information are not likely to be used as much as the genetic marker for the calculation of genomic breeding. Animal models and animal breeding have been provided with new opportunities by the development of genome maps for cattle and poultry, which has already been completed.

Leakey brought forth that currently markers affecting traits or genes cannot be measured easily like in the case of resistance to disease or the quality of meat and the same can be effectively done through DNA-based tests.

However, for the adaption of the livestock to the various changes in climate, genetic resources continue to play an important role. Any challenges in the future can be met with genetic insurance in the form of native breeds.

Genetic diversity in species that are threatened can be maintained through the effective use of modern reproductive technologies whereby stored and frozen germplasm or the genetic resource banks can support the various measures for conservation.

Natural breeding programs can be successfully optimized through the use of modern approaches for monitoring endocrine non-invasively in addition to the direct application of reproductive procedures that are technologically advanced. Farms that collect the entire data electronically and follow close monitoring of that data can have a separate category of progeny test that shall have to be developed accordingly.

Genetic improvement can be benefitted tremendously through the opportunities offered by the parentage verification and collection of data automatically, which shall, in turn, bring about economic benefits.

For identification of the genes that affect various productivity-related parameters, genetic analysis is done by sending the biological samples to the laboratories. Additionally, the requirement for alternative methods can be reduced through selective breeding.

3.3.1. Milking Automation

During the 1990s, the dairy sector involved an automation system for milking. To ensure that illness, quiet heat, normal estrus cycle or stress is recognized safely and quickly in animals, animal husbandry requires objective measuring systems that are suitable.

A completely different management system is required in an automatic milking system for feeding, grazing, milking, cow behaviour and cow traffic which also ensures that the health of animals and quality of milk is safeguarded.

Instead of inspection for abnormality by humans visually, tools like electronic sensors or devices need to take over. However, a basic requirement for developing sensors for abnormal milk detection is the definition of abnormal milk.

The use of sensors in the milking systems is relatively new even though the market has seen their presence for a long time as, during milking, the milkmen would evaluate the milk. The milking system has seen widespread use of sensors as a result of the intelligent milking system's development.

Sensors on the milking robots are capable enough of detecting any signs of mastitis by which in turn several characters of abnormal milk, milk acidity, somatic cell count and milk conductivity among others can be measured. Specific milk specifications in the system like milking efficiency and parlour performances too can be taken into account through these systems.

The rate of milk flow from individual cows can be monitored through automatic cup removal devices that are quite simple, and when a threshold is reached, the system gets activated for the withdrawal of cups from the cows by switching off the milking vacuum. Amongst the many strategies to control mastitis, one of the known components is the disinfection of the teats post-milking.

A dip cup is used more often, and sometimes a pressure operated spray is utilized by farmers to perform this task manually. For monitoring the animals, the milking system is also installed with a behaviour meter. Individual animal's lying bouts, activity, and lying time is continuously recorded by the behaviour meter.

The health and reproductive status of the animal, along with the welfare of the animal in stressful situations and varying environmental conditions, can be gauged through the observations of cow behaviour. Usage of the separation gate at the automatic management systems is another option.

Frequently, the challenging and risky activity of cattle separation needs to be performed. Parlour performances and efficiency of milking reduces where animal separation is done by the milkers also. It would be advantageous to reduce the risk and need for separation of this workforce.

In big herds, the workload of the farmers increases tremendously for the separation and grouping of cattle. Many cattle farms do not commonly use electronic separation gates.

The handling and management operation can be majorly benefitted by the removal of labour that is required for the separation of animals. At the same time, the farmer needs to be made aware of the diseased cows. All the demands in this aspect are not fulfilled currently even though there are some sensors available. Performance suffers when due to the involvement of the operator in the separation of animals; certain other tasks cannot be performed.

It is a major task to identify and draft individuals in the case of larger herds. Many dairy farms are not installed with automatic drafting as a routine. In many fields, farmers are benefitted from the usage of electronic technology. Potentiometric chemical sensors are used in the electronic tongue.

Sensors with chalcogenide glass sensors, plasticized PVC membranes with cross-sensitivity to organic and inorganic anions and cations, glass pH electrode as well as potassium-, sodium-, and chloride selective electrodes are comprised in an array.

Far more effective and faster results are provided by automatic milking systems that use the sensors that have been developed newly (SCC, NIR, and LDH, etc.). For the diagnosis of mastitis, many biosensor-based research studies are still carrying on.

For the detection of mastitis, the current self-contained controller (SCC) and alternative methods were reported by Viguier et al. For the production of good quality milk, many sensors are currently being used. Microchip technologies have yielded faster results.

Furthermore, mastitis can be successfully diagnosed with these technologies where more accurate results are obtained due to the wider range of these tests and results that are more effective. Rapid detection of mastitis is provided through all these results. Even though the appearance of milk and its conductivity is regularly used on the farms, early detection of mastitis that is provided by the other methods can be used for accurate and fast decisions, and that helps to cure the disease.

3.3.2. Feeding Automation

For the preparation of ration, the farmers are offered the best option by the software designed by computer programmers. For advanced options like racing, animal feedstock, live weight and lactation period-related information, optimal feeding programs can be done. The data provided by the National Research Council (NRC) about feed content and animal feed is used by these programs.

For getting better produce, many of the dairy farms that are big use various kinds of automatic systems for feeding the animals. Each stage of feeding right from the preparation of feed, the mixing equipment required to the installations for feed distribution, i.e., the complete system is included in these automated animal feeding systems.

Feed ingredients like corn/maize silage and grass, feed concentrate, and mineral mixture is loaded; thereafter mixed and eventually delivered by the system to the feed table that has been built there.

Simply put the automation systems comprise a programmable command manager, a communication interface, a control panel, a scale, and the entire equipment which is required for organizing the feed provision and feeding process in relation to the animals as per their age groups. As compared to the calf feeding methods that were traditionally used, the calf feeders that are computer-controlled have far more advantages.

The individual calf's daily intake can be monitored through the transponder that calves carry (39). Labour costs reduce tremendously (73%) through this technology and the calves very easily learn to use this computer-controlled milk feeding system. For the welfare of the calves, a health observation system and automatic weighing can be combined with this system.

As compared to calves in hutches, those that are reared in a group pen required medication on far fewer days when calves are fed a milk replacer twice a day from a bucket. Using the Electronic Concentrate Feeding system, it is ensured that, an exact amount of feed is supplied at the right time to each cow.

For the concept of automatic feed supply systems, the ideal introduction is the Belt Feeder feed distributor. A combination of the sliding scraper and conveyor belt, this system is quite small, economical, and flexible.

Various technical approaches to automation have been reported by Grothmann et al. These include the stationer systems like a conveyor belt and rail-guided or self-propelled feeder wagons that are mobile systems.

Amongst the farmers an innovative technique, the rumen activity sensors are quite popular for the reduction of metabolic disorders which is in addition to the automation approaches for the feeding system. To prevent major problems, a farmer can adjust the feed when acidosis is exhibited by the sensitive cows. To find out the rumen temperature and rumen pH, many electronic sensors can be used for the cattle. For 100 continuous days, rumen bolus can work for all future evaluations, and the data is stored on every 15 minutes.

A good indicator for gauging the health condition of cattle is the rumination activity. Rumination gets inhibited by various diseases, stress, and excitement or states of anxiety (66) as the prerequisite for rumination is at least some level of well-being.

Chewing activity can be estimated by collecting data for cow jaw movement for which another sensor is used. When the mouth opens or closes, the changing pressure of the animal cannot be detected and the sensor works on this principle.

3.3.3. Reproductive Performances

For farms that are varying between 13 and 18 months, the best criteria for comparing the reproductive performances are the average calving interval, estrus detection technology, and in most dairy herds, the efficiency for heat detection varies between 30 and 50%.

When inseminated, 5–30% of the cows did not show to be near or in estrus as reflected in the results of research. Many factors like the cow number, statistical method for data analysis, threshold level and barn style influenced the results of estrus detection. False warnings in a large number were indicated through the detection error rates that varied between 17% and 55%.

There should be a charge off of the estrus detection system costs for maintenance and purchase owing to the estrus detection and satisfactory conception rates. Even though all the criteria are not met, some of them can be met by the available technologies and aids that are of both kinds inexpensive as well as expensive ones.

In many countries, visual observation of the dairy herd is done for the detection of estrus even though in large dairy farms, this procedure is very difficult as the observation periods during milking and feeding are quite short. As reported by Galiç et al. in comparison to large farms, the success rate in the smaller ones is higher. The size of the herd has a significant effect ($P < 0.0$) on the milk yield, lactation number, calving age and calving interval.

With a standard deviation of 2.3 h, Schofield et al. calculated the mean duration of estrus to be 13.5 h. It has become possible to detect estrus automatically due to the technical progress so that cows can be monitored through computers. There has been an analysis of varied traits in many studies so that they can be utilized for the automatic detection of estrus.

A computerized estrus detection device is the radio telemetry whereby through the electronic system the device can detect electronically as to whether a cow needs to be mounted by a herd male followed by monitoring of the activity continuously.

The mounting activity of the cows can also be used in other possibility-the patches. A signal is sent to the receiver by the depressing of the transmitter when a cow is mounted by another cow. The main computer receives the duration, time, and date of the mount during this time which is stored and sent to it. For a final decision, this entire data is prepared after evaluation on the computer.

As compared to other methods, even though the related costs are high, the accuracy of detecting estrous increases in this computerized detection of oestrous so; they are paid off by the benefits. Huge losses may occurs to the dairy farms due to errors in the detection of oestrus.

For dairy farms, the economic losses can vary from $2 to $6 per day. However, for a cow, one cycle missed means a cost of $42 to $126. These losses can be prevented through the usage of detection aids which becomes advantageous.

Pedometers compare the previous activity data by storing the current and past physical activity and can be used for the detection of estrus. The report is prepared after analyzing the data programs for active cows which is then accepted as oestrus. Farmers are alerted through a flashing light or beeper that the cow needs to be controlled.

The cow's activity over a unit time can also be measured through the pedometers that are attached to the leg of the cow for detecting oestrus.

The market has a number of commercially available pedometric systems in addition to the standing activity systems. The mounting cow activates standing activity system. A personal computer receives the data for analysis from the receiver that relays to the buffer the radio signal that it has picked up. The number of cows, date, standing time and the duration is recorded by the system so that evaluation can be done on time.

According to Chung et al. as a standalone solution, voice identification processing can be used where accurate (with an accuracy of 94%) and economic (simple microphone is used) detection of estrus can be done. The Heat Watch system's Mount Count manual version is available at a very low price in the market, and no software or hardware is required for the processing and display of data in this.

Also, a pressure-sensitive device can be mounted on each cow's back so that when the cow stands to mount, the device gets triggered. The programming of the pressure-sensitive device is done in a manner that when several mounts have been recorded that are valid, signals are given by a light. Whether or not the standing estrus is a teaser or a marker animal can be effectively detected through the second one.

Marking devices are put on the marker animals. Marks are left on the rump and back of the animal, which is standing in estrus once the marker animal has mounted due to the rubbing of the chin-ball marker against the animal in standing estrus.

Estrus can be effectively detected through standing and mounting activity. A series of other methods like cervical mucus, temperature, hormone changes in the milk and blood, vaginal characteristics and blood flow are also available on the system.

However, at the farm level, these methods cannot be applied. The cycle or pregnancy's stage can be deduced through the criteria of progesterone levels in the milk. This is a very good criterion and problem cows in a herd can be diagnosed through this.

For many purposes, the behaviour of the animals like activity, lying bouts, and lying time for each cow can be recorded continuously by the behaviour meter. The health and reproductive status, as well as the welfare of the animals, can be assessed through the cow's behaviour in various stress and environmental conditions.

Pregnancy check: to get an ideal calving interval, one of the most important factors is the diagnosis of pregnancy. Transrectal and rectal ultrasonography of the reproductive tract are the most common methods.

Pregnancy can be diagnosed reliably from 35 days by a veterinarian who is experienced and from 30 days by a practitioner who is experienced. A marker of pregnancy can be detected by using either milk or blood in tests like latex agglutination (LA), radioimmunoassay (RIA) or Enzyme-linked immunosorbent assay (ELISA).

Pregnancy in cattle can be indicating by glycoprotein, progesterone, and estrone sulfate. High specificity is offered by estrone sulfate as it is produced by the fetus, where it is essentially a conjugated steroid product of estrone. The main drawback of the test is that pregnancies cannot be diagnosed reliably before hundreds of days of gestation, and the rate of false negatives is quite high. In multiple cows, early stages of pregnancy have been detected by the measurement of many characteristics through a wireless system designed for the same.

3.4. METHODS OF ANIMAL BREEDING

Based on the breeding work done in cattle, out-breeding, and in-breeding are the two methods of animal breeding (Figure 3.1).

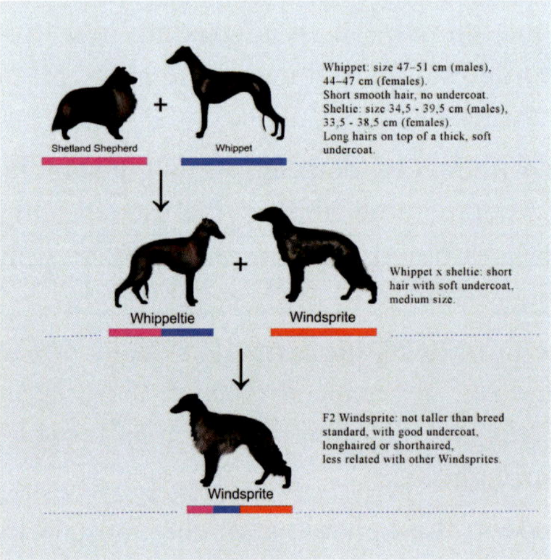

Figure 3.1. Dog 2-generation outcross graph.

Source: *Image by Wikimedia Commons.*

3.4.1. Out Breeding

Sometimes animals from different breeds (cross-breeding), different species (interspecific hybridization) or the same breed (but with different ancestors) are bred together where they are unrelated to each other and this entails outbreeding:

1. **Out Crossing:** When no common ancestors exist up to four to six generations of the pedigree on either side of the mating animals which are of the same breed it is outcrossing. Outcross is the resulting offspring from such a cross. Outcrossing works best for animals that have productivity that is below average in the production of milk, with respect to the growth rate in beef

cattle, etc., sometimes inbreeding depression can be dealt with by a single outcross.

2. **Cross-Breeding:** Superior females of one breed are mated with superior males of another breed in cross-breeding. Better breeds are produced through this method, and many new breeds have been developed over a while via this strategy. Better progeny can be obtained by mating superior bulls with cows that are of an inferior breed. In Punjab Marino rams have been crossed with Bikaneri ewes to produce a new breed of sheep, the Hisardale.
3. **Interspecific Hybridization:** Mating of female and male animals of two species that are different is done by interspecific hybridization. This kind of a mating produces progeny that is in variation with both the species that were its parents.

Sometimes, both the parent's desirable characters get combined in the progeny. For instance, the cross between a male donkey and a mare (female horse) produces a mule. Mules are better suited for mountainous regions and can work hard, in fact much harder than both their parents.

4. **Grading up:** Grading up is the practice of breeding sires of a given breed to non-descript females and their offspring for generation to generation. It is the succesive use of purebred bulls of a certain breed of non-purebred herds. The continued use of good purebred sires for only few generations are all that are required to bring the herd to the point at which it has all the appearance, actions and practical values of pure breeds.

3.4.2. Inbreeding

When for four to six generations breeding is done between animals of the same breed, it entails in-breeding. Bulls and cows can be taken as examples to explain inbreeding. Identification and mating are done of cows and bulls that are superior to those of the same breed. Evaluation is done of the progeny thus obtained from such mating, and identification of the superior ones is done for mating further. For instance, in cattle, a cow producing more milk per lactation is a superior female. Whereas a bull is a superior male when compared to the other males the progeny raised by it is superior. Hence if a pure line is intended to be developed in any animal, inbreeding becomes necessary. Recessive genes exposed by inbreeding can be eliminated through selection. At the same time, the less desirable genes are eliminated, and the superior ones are accumulated. However, productivity and fertility both reduce due to inbreeding.

This is termed inbreeding depression. Yield and fertility can be restored in this kind of a condition by breeding selected animals from the population and mating them with animals with which they are unrelated but are superior to those selected animals.

3.5. CONTROLLED BREEDING EXPERIMENTS

Multiple ovulation embryo transfer technology (MOET) and AI are used to carry out controlled breeding experiments:

1. **Artificial Insemination (AI):** The reproductive tract of a selected female is injected with the semen collected from a superior male by the breeder. The semen may be frozen for usage at a future point in time, or it may be used immediately.

About 5 to 10 billion sperms get deposited in the vagina of a cow when the cow is inseminated by the bull naturally. On the other hand, to achieve conception, a far lesser number of sperms are required when these are artificially deposited. Hence AI turns out to be a far more economical process. At the same time, this method can control certain diseases from spreading.

2. **Multiple Ovulation Embryo Transfer Technology (MOET):** Superovulation and follicular maturation can be induced in the cow by giving it hormones (which have activity similar to FSH) whereby 6–8 eggs are produced per cycle instead of one egg. AI is done or the best bull is used to mate with the cow.

When embryos are at the 8–32 cell stage, they are recovered and placed into the surrogate mothers. Superovulation can then again be performed with the genetic mother. Cattle, buffaloes, rabbits, mares, sheep, etc., have been successfully bred through MOET. In a short time, better breeds have been produced by the successful mating of high-quality meat (where the meat is lean and has fewer lipids), giving bulls and females of the breeds that give high milk.

3.6. TECHNOLOGIES THAT IMPROVE PRODUCTIVITY IN THE LIVESTOCK INDUSTRY

The last few years have seen a lot of changes in the livestock industry. There is a clear purpose to the new technologies being used in the livestock industry where the main role is played by automation. The purpose here is to make the life of the farmer easier, improve the productivity of the livestock business and at the same time make the best use of the available natural resources. Main livestock industry technologies aimed at improving production efficiency (Figure 3.2).

Figure 3.2. Technologies in cows for better management.

Source: *Image by DataMuster.*

3.6.1. Ventilation Adjustment Systems in Farms

Farms now have far more evolved air conditioning, unlike the fans that worked in the past. The aim now is to enhance the well-being of the animals by improving the environment of the farm. Both for the farming and livestock industry, an important project has been developed by common language runtime (CLR) for a national company whereby the electronic control of the ventilation system is targeted.

The design of the hypocycloid project was involved in this project so that it could be adapted to the tubular apace where it was required to be housed. A system of sensors that are controlled automatically operates the displacement systems by which the farm ventilation is governed.

3.6.2. Automated Camera

The automated camera is connected directly to the smartphone, tablet, or computer of the farmer, revealing relevant warnings, and it can operate under different kinds of light conditions. In this system, valuable information about the animals and their condition with respect to the liquid and solid nourishment is relayed through the probes that are placed below the trough from which the animals feed (Figure 3.3).

Figure 3.3. Livestock monitoring cameras.

Source: *Image by West Country Tech.*

3.6.3. LED Lighting Technology for Farms

The service life of the CCF fluorescent tubes and CFL lamps that save a lot of energy is relatively lower compared to the other solutions, yet they are looked up to in industrial farms. The usage of LED lighting has seen a major escalation due to their high energy-saving capabilities.

Their main advantage is that no hazardous waste is generated by them, and at the same time when their durability is measured in hours, they are extremely beneficial. The LM/W ratio of these lighting systems should be known so that maximum efficiency can be derived wherein the higher the ratio, the better it is.

3.6.4. Drones for Grazing Monitoring

In livestock production for monitoring grazing, the usage of drones in the livestock industry is a very innovative method. The drones can capture images through the use of multi-spectrum cameras and infrared sensors whereby aerial capturing of images that are quite interesting can be done so that the health condition of the animal, biomass, and the grazing amongst the animal population along with the other factors in livestock production can be monitored (Figure 3.4).

Figure 3.4. Drones for monitoring cows.

Source: *Image by AgUpdate.*

3.6.5. Automatic Incubators for Poultry Farms

CLR is involved in this technological aspect of automatic incubators as well. To ensure that the eggs are properly incubated, proper rotation at the correct speed is allowed off the incubator machines through gear-motor-operated panels in exactly the same manner as a hen would. This ensures the maintenance of the right temperature and heat.

3.6.6. Automatic Milking Robots

Up to 70 cows can be controlled and milked by milking robots that are automated. The animal is milked whilst it feeds on ration through a mechanical arm that is controlled by sensors. The personal file of each cow can be built by new robots so that their age, origin, health condition and milk quality can be controlled. Using robots for this purpose not only reduces the labour costs but also helps to increase the yield of milk (Figure 3.5).

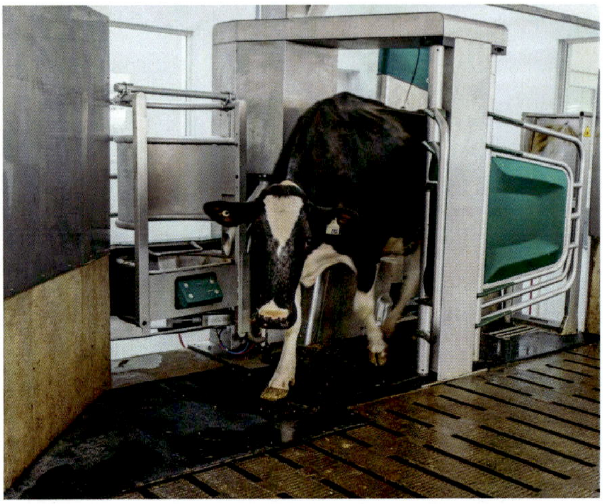

Figure 3.5. Robotic milking system for cows.

Source: *Image by AgUpdate.*

3.7. CLR: ELECTROMECHANICAL ENGINEERING FOR IMPROVING THE LIVESTOCK INDUSTRY

Execution of multiple innovative projects in the livestock industry sees active participation from CLR as it keeps itself updated about the automation process and modernization that is taking place in this field. Cost-effectiveness along with high performance is offered to the producers through the gear motors of the highest quality being developed by the engineering team of CLR whereby the strictest standards are complied with for achievement of solutions.

For more than two decades, various applications in the livestock industry have seen the presence of various electromechanical applications of CLR like in incubators, closure, and aperture of rationing feed troughs, incubators, etc.

The rapid digitalization process and modernization is making its presence felt in this industry. As a result of this, for operations of different sizes, the focus of the new technologies is on making the farming activity a viable sector.

CLR invites outsiders to look at their free resource that is focused on suppliers of actuator solutions and selection of gear motors and which for those in the field of electromechanical engineering is of great interest.

3.8. NOVEL TECHNOLOGIES IN ANIMAL HUSBANDRY

Records are showing that during the Harappan times, a number of species of barley, rice, and wheat were found. Field peas and other varieties of leguminous plants, dates, sesame, and melons too were grown around that time. Mustard seeds have been found at Chanhu-Jo-Daro. There is evidence to show that cotton too was grown during this period from which textiles were made.

At the same time, the excavation sites of the cities at Harappa throw light on the domestication of numerous animals like the humped cattle, buffalo, goats, ass, *Bos indicus*, *B. bubalis* sheep and camel. From the indigenous jungle fowl, domestic fowl was domesticated, as can be seen from the presence of their bones at these sites.

Even though dogs and cats too were tamed, the presence of elephant bones does not confirm whether the latter was domesticated. It was only during the Harappan period's last stages that the horse made its presence felt even though its presence was quite rare.

Livestock holds an important position in any society that is agrarian. A few of the products in use in such a society are milk, eggs, meat, and hide. The population has been increasing rapidly but to overcome the discouraging state of affairs in this sector, efforts are being made to bring in technology related to animal products so that the production can be potentially useful and beneficial to humans.

The living standards of the common man have been greatly improved upon since the onset of the industrial revolution, which brought about a series of attractive and useful applications. New technology has benefitted in many ways like increasing the cost efficiency, improved working conditions, improved provision of important production data; better monitoring (for instance, access to real-time data and monitoring remotely) and enhanced welfare of the animals.

Above all, technology is seeing rapid development. An important contribution towards this development has been through the transfer of software and computer systems to the application.

Animal production can be increased efficiently and profitable manner whilst the working conditions of the farmers are made comfortable through various technologic instruments. Hence the main research area for animal sustainability and productivity lies in the field of technological developments.

Animal husbandry has been made comfortable and much easier through various technological tools and equipment. With rapid developments in these applications and management decisions have been highly affected.

As per the correctness of the decisions that are to be made, management decisions need to be configured and made daily in animal husbandry. Farmers are provided with a tremendous number of opportunities through smart systems.

Correct Management decisions affect daily jobs like feeding, milking, reproductive performance, and environmental control. Massive repercussions take place as far as the profitability and quality of the product are concerned when the decisions made and work done is hampered by human errors.

Important information concerning the challenges being currently faced as well as the key innovations by which animal husbandry is being affected is dealt with in this chapter. Further, with detailed research results, innovative applications and approaches with respect to animal husbandry have been summarized (SerapGöncü). A decade ago, Thornton and his co-workers brought out a very basic problem that was affecting livestock production. The number of animals and their production per farm was seeing a steady rise even in light of the growing population and its demands for quality products even though the size of the farms continued to be the same, thus giving rise to problems in the livestock production.

It has been proven through various researches that a multidisciplinary approach and study is essential for a resolution to these problems, and the usage of different technologies can be of help.

For large enterprises to show an optimal performance, the usage of automation systems and technology is an essential requirement, especially when it comes to high genetic value animals. On a day-to-day basis, the work at a livestock farm is quite simple and standard application of procedures is done. The production can be controlled, and animal health, as well as overall welfare, can be looked into in the modern dairy farms through data monitoring.

The work of the producers gets easier through the use of new technology so that an improvement can be brought about in the welfare of cattle, profitability, and efficiency in production. Farmers can get the on-time process through the fast, profitable, and efficient solutions that technology provides through the possibilities created in direct manipulation and management. The management of animals needs to ensure their well-being which can be brought about through careful management and monitoring of disease continuously.

Animal husbandry and realizing its importance is the cornerstone for the application of technology in animal husbandry and the management of livestock. Animal husbandry entails the maximization of benefits that accrue from housing, breeding, feeding, and ensuring the health care of livestock. When domesticated or farm animals like sheep, cow, etc., are reared by humans for commercial uses, they come in the category of livestock.

When animals are reared for profit purposes through the application of technology, it leads to animal production, which basically comprises housing, breeding, feeding as well as their marketing. To give the animals a good and quality life scientist do in-depth study of the various factors which influence the rearing of the animals along with how their products are marketed and manufactured.

Domesticated animals brought up to produce commodities like eggs, milk, meat, leather, fur, and wool become livestock. Usually, livestock refers to animals that are raised for consumption; on the other hand, it also refers to farmed ruminants like goats and cattle. All over the world, agricultural lands' the largest user is livestock production.

Livestock related technology is very vast. During the last 50 years, meat production has trebled. Further rising income levels have affected consumption bringing about changes in dietary patterns. This has only resulted in the need for farmers to produce more even though the resources are fewer. Moreover, the diversified use of the new technologies can do a lot for the managers of livestock production.

3.9. USE OF IOT IN ANIMAL HUSBANDRY

3.9.1. Monitoring of Metabolites

It is important to monitor the metabolic activities of animals. For the technology to be apt for animals, bio-nano sensors are being integrated so that they can grasp the metabolites like glucose, adenosine triphosphate (ATP), and lactate. Electrochemical additives that are mixed with an antenna having radio frequency verbal exchange gadget are contained in these sensors.

When this system was examined on models of animals, it proved dependable. When these bio-nano sensors were introduced into the animal models they were also tested for tissue infection. Via suture, inside the subcutaneous environment, the attention of ATP did not show any impact. However, through invasive procedures like puncturing for implantation of Nano-Bio-Sensors or biochip neutrophils infiltration was detected.

In a haemorrhage, the intestinal tract is involved whereby the organs suffer an ischemic injury whereby using lactate and glucose the injury can be gauged. A variant of intracellular biochemical markers causes intestinal ischemia through the aid of which the bowel beneath strain can be tracked physiologically. The effect that hypoxia has on the metabolites have been displayed efficiently by using online biosensors.

It can be quite a difficult task to track the metabolism of animals that are unfastened and are being transferred. There has been a proposal for the usage of a tool that can be implanted where it has a micro-fabricated sensing platform, is custom-designed with circuits and has a coil of electricity.

The exogenous and endogenous metabolites can be detected electrochemically through this. For biocompatibility, a polyurethane membrane that is epoxy-stronger with a capability of retaining enzyme interest for about 35 days was used.

On the web page of the implanted device, less inflammation is produced when the biocompatible membrane is used. Even though the extra prototypes of the electrochemical sensors for lactate and glucose size in bodies of animals are being verified, these are available commercially. The medicine dosages for animals may be customized by the usage of implantable devices that can monitor drugs.

For monitoring the numerous metabolites healing range and a fast analytical detection electrochemical sensor are being proved to be the ideal gadgets. Nano-structured substances are used to progress the sensitivity of the sensors. At the same time, the sensors' biocompatibility should be looked into.

Even though carbon nano-tubes have worked well in researches conducted recently, they shall be toxic. To ensure that the host animal accepts the sensor and shows minimal reaction to the foreign object, the sensor balance needs to be borne in mind whilst the animal frame is being implanted with a sensor. Also, heavy batteries are required as sufficient electricity is required for the gadgets, which may result in the device becoming undesirably bulky.

3.9.2. Breath Analyzer

The stage of blood glucose can be gauged in-depth by an analysis of the volatile natural compounds that are present in the breath. For this reason, there is a separation of the volatile organic compounds or the VOCs. The blood glucose levels relate to the VOCs like the body's ethanol, ketone, exogenous mixes, and methanol.

Moreover, for breath analysis, the exhaled breath's risky composition can be utilized, and for urgent diagnosis, this is a noninvasive method. A series of chronic respiration and cardiovascular illnesses comprise these illnesses. The organism's metabolism can be completely read through the display of airways and bloodstream's composition which is revealed in the breath's unstable composition.

Modern analytical technologies can be added to the methods like needle trap and stable-segment microextraction along with mass spectrometry which allows the exhaled breath to be analyzed. An early diagnosis of the disorder can then be done by evaluating received heterogeneous datasets statistically. Disease detection can be done through digital gadgets that detect the heady scent from animals.

Tuberculosis and most cancers of the lung are detected in rats and puppies using this method. The presence of certain kinds of volatile compounds can be detected due to the special sensory cells that their noses have. This complicated digital device has been attempted to be copied in digital devices that are made by man like Enosis albeit with little achievement.

A livestock sickness, namely bovine tuberculosis has worldwide significance for public health. Pathogens can be used to discover the produced volatile natural compounds, and this has garnered tremendous interest for finding remedies in both veterinary as well as humans for the diagnosis of this illness. The profiles of the unstable organic compound that are present in breath can easily help to recognize the contamination by M. Bovis.

A lot of snared breath biomarkers which are similar to methane (CH_4), carbon dioxide (CO_2), alkali, and nitric oxide can be investigated by spectroscopic procedures with high affectability like consolidated depression yield spectroscopy (ICOS), pit better assimilation spectroscopy (CEAS), as tunable diode laser ingestion spectroscopy (TDLAS), quartz unrivalled photoacoustic spectroscopy (QEPAS) and), dig space spill out spectroscopy (CALOS).

The biomarkers' fingerprints extend from UV to IR region. Currently, the utilization of these sensors and the laser spectroscopic systems is accessible and is utilized industrially for the examination of breadth. Concerning to the M. Bovis disease, the nearness of VOCs related to it can be screened through the mass spectrometry/gas chromatography examination.

Homestead animals that have been affected by M. Bovis can be identified by analyzing their breath through the exhibit of sensors that are based on nano-technology through which cows can be observed in genuine time. A non-invasive screening method that is quite popular can be used to recognize the foot-and-mouth ailment's source.

Airborne infectious sellers are captured by the hand-held air samplers employed by this technique which uses electrostatic particle size which is then subjected to a polymerase chain reaction (PCR) in actual time. In contingencies of the epidemiological kind, the virus for foot and mouth disease can be monitored.

With pattern popularity techniques, the array of sensors that have monolayer-capped gold nanoparticle (GNP) can be successfully used for powerful and instant diagnostic consequences from the samples of exhaled breath. Monolayer-capped GNP sensors can be utilized for breath analysis with the intent of deriving benefits in extraordinary fields of medicine like oncology, respiratory medicinal drug and infect ology, and this has been fairly well established. The CO_2 from the fluid can be targeted through a recent patent through which the respiratory health o the animals can be determined beneficially.

3.9.3. Virtual Animal Fitness

The animal fitness market is being revolutionized by advances in the digital era. The connection between pet proprietors and veterinarians has been reinforced through the development of many applications. All leading businesses have engineers who are developing virtual generation so that the desires of dairy farmers, puppy proprietors and cattle managers can be met.

The dietary intake, socialization, exercise, and other activities can be monitored through an application called the PetDialog where alerts for routine matters of care, including vaccine schedule are sent as alerts through an integrated calendar.

New developments in the digital system have made it easier and simpler to control the fitness of animals, whereas earlier, online appointment scheduling was missed at veterinary hospitals, and bulky procedures for covering animals was involved along with a complicated-on line cash transfer system.

For medical aid that is faster and correct, smartphone-based applications can record music and send animal behaviour to the veterinarians. The practices in food manufacturing that are long-standing can be automated through specially designed applications.

BCS condition an application specially designed to the song a cow's frame conditions has been released in 10 languages due to this capability. Many human illnesses have their origins in animals; hence the outbreak and spread of these illnesses can be controlled and manipulated with the assistance of virtual fitness technologies.

3.9.4. Financial Results

During the last 10 years, a series of resourceful concepts and sensing gadgets have been demonstrated as well as proposed all the same there still seems to be a bottleneck as far as the construction of a miniaturized device is concerned through which in real-time data can be transmitted and simultaneously cover multiple-goal molecules.

When farmers invest in sensors, they need to evaluate the monetary dimensions involved in it. The impact that the usage of sensor structures has had and the actual productiveness of the farms can be deduced through the farm's accounting records. For instance, the monetary results that these investments yielded have been clearly proven in the case of 217 Dutch dairy farms.

Whether or not the farms have sensor systems the productiveness alternate for farms can be measured through the Malmquist total component productivity index. The gradual adjustments in a farm's productivity are exhibited through this index which reveals how far the success of the farm has been contributed to through technical modifications.

For the deployment of sensors that is economically viable in dairy farms, it is essential to explore the effect these systems have on productivity as the latest studies have shown them to greatly influence productivity. For further improvement of the sensor systems' performance, approximate strategies should be briefed to the farmer so that the claims of the producers too can materialize on the actual farms.

3.10. CONCLUSION

To sum up, several techniques are prevalent or in the development phase that can boost the overall operations in animal husbandry. These techniques make the routine activities of the people easy who deal with the management of dairy animals. The intended aim of all these techniques is to do most of the tasks effectively and efficiently

However, it is expected that with an increase in the demand of people on animals, there expected demand will likely increase in future, which in turn through a challenge in this industry. Therefore, future technology should take into consideration all these factors and others factors while developing new technology for animal husbandry.

In addition, some of these techniques are expensive to employ, therefore, farmers, especially in developing countries resist applying them to their dairy farm. Thus, it will be better if there is low cost so that it can be affordable to all, which will further help in the growth and development of this sector on a large scale.

REFERENCES

1. Banerjee, G.C. (1999). *A textbook of Animal Husbandry*. 8th edn., Oxford & IBH publishing Co. Pvt. Ltd., New Delhi, India, pp-1079.
2. Biology Discussion, (n.d). *Animal Breeding: Objective and Methods of Animal Breeding*. [Online] Available at: https://www.biologydiscussion.com/breeding/animal-breeding-objective-and-methods-of-animal-breeding/1369 (accessed on 15 March 2021).
3. Göncü, S., & Güngör, C., (2018). The innovative techniques in animal husbandry. *Animal Husbandry and Nutrition*. [Online] Available at: https://www.intechopen.com/books/animal-husbandry-and-nutrition/the-innovative-techniques-in-animal-husbandry (accessed on 15 March 2021).
4. International Livestock, H., (2021). *Technologies that Improve Productivity in the Livestock Industry*. [Online] Clr.es. Available at: https://clr.es/blog/livestock-industry-technologies/ (accessed on 15 March 2021).
5. First, N. L. (1990). *New Animal Breeding Techniques and Their Application*. [Online] PubMed. Available at: https://pubmed.ncbi.nlm.nih.gov/2213713/ (accessed on 15 March 2021).
6. Wankhede, D., & Pathakala, M., (n.d). Use of IoT in animal husbandry. [eBook] *IOSR Journal of Computer Engineering (IOSR-JCE)*, pp. e-ISSN: 2278-0661, p-ISSN: 2278-8727. Available at: http://www.iosrjournals.org/iosr-jce/papers/Conf.17025-2017/Volume-2/7.%2040-44.pdf (accessed on 15 March 2021).

Chapter 4

History and Importance of Animal Nutrition

CONTENTS

Animal husbandry is a part of science that is concerned with raising animals. In this chapter, the history of animal husbandry and nutritional requirements for animals are discussed. The first part of the chapter discusses the Archaeological approaches to early animal domestication: concepts and techniques and their limitations. Followed by the early domestication of animals on a global scale, the chapter gives meaningful insights on animal nutrition and its requirements in livestock. The chapter ends with a discussion regarding the meaning and the history of domestication.

4.1. INTRODUCTION

The first species known to have been domesticated are Wolves (Canis lupus). European and Asian hunter-gatherers had domesticated it during the Late Glacial. There were no major modifications in the way of life of humans. There were only small changes in the hunting strategies, tactics, or techniques. As per the observations, it is clear, that the domestication of Homo sapiens was already there during the Upper Palaeolithic. However, this was utilized for only a restricted number of species in certain circumstances. In other parts of the world, there were several other species of animals and plants that began to be domesticated. The domestication of fish is still a primary activity in several parts of the world. However, few species have been domesticated of all the biodiversity present in the world. Till present, there is a continuation in the domestication of the new species. In contrary to the domestication of dogs, there was an increase in the number of human societies in the world. This was the result of the domestication of several species in the world. This is the process called neolithization. This process is characterized by deep social and spiritual change along with a strong demographic transition. There is a drastic techno-economic shift starting from hunting-gathering to food production. However, this process was a bit slow. This was based on the cultivation and husbandry of domestic animals.

4.2. ARCHAEOLOGICAL APPROACHES TO EARLY ANIMAL DOMESTICATION: CONCEPTS AND TECHNIQUES

They can be dated precisely if they come from a well-characterized as well as well-dated archaeological context (± some decades to 2–3 centuries). The process of domestication can be well analyzed with the use of high temporal resolution. This is done by direct radiocarbon dating of the collagen from the bones themselves. This analysis can be done for even the early period Ca. 12–10 kyrs before present (BP).

The paleontological or even Pleistocene collections often contains a large series of quantitative approaches that helps in observing the statistical appreciation.

Holocene archaeological bones and Late Glacial are not considered as fossils. There is well preservation of the histological structures, associated unicellular or helminthic parasites and organic matters. This allows the analysts to deeply analyze the biological aspects of it that includes paleomolecular or isotopic approaches.

There is a very large and informative panel of pre-mortem (or intrinsic) biological signatures. They aim at providing information on the size and shape along with multiscale and refined analysis. It is done using geometric morphometrics. Geometric morphometrics gives information on the evolution of the size and shape and the age patterns of the animals.

The relationship of the animal species with that of humans is indicated by the cut marks and the cooking burns on the body of the animals. This is also indicated by the archaeological context of discovery (food refuse deposits, human burial, cultural sites).

4.2.1. Conversely Archaeological Approaches Are Limited

Archaeozoological discoveries are not yet studied and, also not documented well. It comes from limited regions and periods.

The archaeological analyzes are done at a low pace. The animal bones cannot be used unless there are 5 to 10 yearly excavation sessions before the contextual information is available in the refined chronological order.

There is much biological information that is lost. The thing reserved is the information relevant to skeleton.

The quality and the quantity of the information relevant to archaeozoology is reduced due to extreme fragmentation of the bones. This fragmentation is caused by the systematic consumption of marrow. This is also due to the post-depositional attrition of the collections.

Therefore, the in-depth knowledge of both the archaeological context and the taphonomic process is of utmost importance. Since it has degraded the information, it is important to analyze the faunal collections that are archaeologically preserved. This also includes paleomolecular or isotopic analysis.

For this, there must exist a collaboration that is well-balanced and tight among the excavator the osteo-archaeologist and the specialist. The excavator is one whose scientific approach should be e as difficult and important as that of the analyst. The osteo-archaeologist is the one who is the in-charge of the faunal assemblages and its general study. The specialist is the one who undertakes the investigations relevant to the molecular geomorphometric and isotopic.

To understand early Neolithic domestication, there must be a clear theoretical view of what domestication means. As per archaeologists, domestication is the process of the reproduction of a deme. Deme is the local sub-population of animals or plants.

Human society uses the reproduction of animals or plants for their social material and symbolic profit. Here in this definition of domestication, it is clear, that it is very different from that keeping of some Amazonian, Japanese for New Guinean Ainu hunter-gatherers.

As they can capture the young wild animal for a particular house instead of a whole society. They give a symbolic offering to nature by later releasing for killing the animal without having any ok spring to guarantee their future subsistence.

There are various forms of domestication which is the result of the multiplicity of human behaviour. Since it is a process dependent on the animals or plants species, it may take many forms.

The mutualistic relationships between animal and human societies are helpful in the arrangement of the gradient of eco-anthropology. The relationship has extended from anthropophilic to commensalisms. In other words, it has extended to control in the wild, managing the captive animals, expansive or intensive breeding, and ultimately to pets.

It is always possible to achieve sustained stability at any level, as the process only depends on a dynamic equilibrium between animals and humans. This is going to be done with for the progression or getting back to unless intensive relationship is possible. The situation is clearly illustrated in today's time due to diverse situations obtained and by the phenomenon of feralization.

The question arises that from which step did the domestication began. This is the main deciding factor in archaeology. From this, what comes in the frame of domestication is the control of the captive, while commensalisms and control in the wild cannot be considered as domestication. While the control of captives can be considered as domestication as per the degree of animal reproductive control.

Domestication has its own effects on plants and animals, including biological effect. These effects must be studied and analyzed apart. The biological effect of domestication is different in different kind of species.

There are several Constant modifications like increased fertility, decrease in aggressiveness, sexual dimorphism, body size decrease, shortening of the face, a decrease of the braincase volume, the appearance of new coat colours, soft ears and voice changing in case of dogs and other vertebrates including birds and fish.

These biological changes may occur with or without human intent. Some of the modifications can be the result of hormonal changes in animals, or they may also occur due to environmental conditions or the stress of captivity, i.e., without human intent. Nowadays, the possibility of selecting modifications by humans has increased, which are purely due to genetic mutations, which in return develop some epigenetics changes in the animals.

Some zoologist believes that the breeding of captive animal leads to these biological signatures of domestication which appeared almost immediately after the breeding of captive animals begin.

With these facts in mind, the main concern of a zoologist is to describe the characteristic of a particular situation, as accurately as possible. The type of modification, the species, and the intensity, along with the nature of the relationships that animals have, are the main factors on which the zoologist can analyze the situations and the biological signatures of domestication can be studied.

This theoretical framework suggests that the concept and process of domestication must be separated from that of domestic animals. To recognize the early domestic animal and the early domestication, archaeozoology has to address the two-concept connection by combining the large panel of information of the remains of archaeological skeletal, but mutually exclusive, questions such as:

- The description of the level of domestication (intensity and nature of the relation); and
- The biological modifications due to domestication.

The first part can be answered by analyzing the presence of animals in the burials by humans and the presence of a species out of its natural area of distribution. It can be examined by obtaining information about subsistence specialization and the content of the high frequency of taxa in the food refuse.

The other being the frequency of the skeletal parts, which can tell the distance between the slaughtering and dwelling places. The combination of sex ratio and the frequency of ages at the time of death are also known as 'paleodemographic' data.

4.3. EARLY ANIMAL DOMESTICATION ON A GLOBAL SCALE

From the available information and data, a brief report about the world's earliest mammal domestications will provide the opportunity to draw out the chief features of this event at a comprehensive scale.

Except for the dog, as discussed above, the earliest perceived domestications are from the Near East. They are basically the Oriental mouflon (Ovis Orientalis), which was believed to have given birth to sheep (*O. aries*), extinct aurochs (*Bos primigenius*), that produced domestic cattle (*Bos taurus*), bezoar goat, which is believed to be the ancestor of the domestic goat (*Capra hircus*), and wild boar (*Sus scrofa*), the domestic form of which being the pig (*Sus domesticus*).

The domestications date of all these are roughly from the middle of the 11th millennium BP. This comprises the cat (*Felis silvestris*), emerging domestication that is expected to concur with the early development of agricultural stock keeping and the succeeding growth of commensal mice.

Although, Goats were believed to be domesticated far later in the time (ca. 10 kyrs BP) in the Zagros by making use of several local distinctive lineages. Domestication of Cattle also initiated for a 2nd time (ca. 8.5 kyrs BP) in the lower Indus Valley (Pakistan) with a distinctive lineage, known as the humped Asian subspecies of aurochs (*Bos p. namadicus*).

The presence of a 3rd autonomous support of cattle domestication in the high Nile valley is still the topic that is discussed widely. Another autonomous centre of pig domestication has been demonstrated in China (ca. 8,000 BP).

Several earlier evidence that are available to date proposed that only the pig (and possibly the dog) had been domesticated locally here. Although several examination and interpretation on the modern genetic lineages of domestic and wild Sus showed various novel potential centres of pig domestication in the region of Southeast Asia, though presently, there is no consistent archaeological evidence that can validate or give weight to this statement.

In Europe, local wild boar domestications have also been evidenced on the basis of genetic and osteological data, ca. 7–5th millennia BP. During a similar period of time and alpaca (L. pacos) and (7th mill. BP) llama (*Lama Glama*) were domesticated in Peru, from L. vicugna, and L. guanicoe, which were hybridized at the later time.

Another wave of large mammal domestication was arose in the Old World during the time period of 6–4th millennia BP. Therefore, it is right to assume that horse (Equus caballus) was domesticated ca. 5.5 kyrs BP in Kazakhstan and probably also during the 5th millennium BP in the Iberian Peninsula.

The origins of camels (*Camelus dromedaries*, *Camelus bactrianus*) and ass (*Equus asinus*) are still being the most debatable topic, but it appears that, based on the available evidence, they had not been domesticated before the 6th and 4th millennia BP, correspondingly.

Finally, the earliest evidence for the domestic water buffalo (known as *Bubalus Bubalus*) has been discovered in the region of Indus Valley, dating to the 5th millennium PB. While, the domestication of birds (Barbary duck and turkey in America, and hen in South-East Asia) came at a later period.

Two of the major trends of early domestication have been pointed out in the present overview. The early domestication event has been contemporaneous with the main climatic crises. For instance, it was ca. 12,500 BP 4th last cold face of the younger dryas, for the increased Holocene warming, ca. 11,500 BP and for the Holocene cooling. It was observed that the domestication of all large mammals took place during the Holocene. Holocene is of the period that provided favourable climatic condition and which played a most important role in the process of domestication. However, some process of domestication has been initiated during the period of Younger Dryas. It is also observed that some of the domestications, such as pigs in China Llama in America, began independently. However, it could have been started by contact with other areas where domestication was already been practised. This shows that the similar practice in the different parts of the world was a doctor at precisely the same time and most probably under similar climatic conditions within the same relatively short period of time. These major and unique changes only to be understood, which requires the study on a much smaller scale.

4.4. ANIMAL NUTRITION

Nutritional requirements are equally important in animals. It helps in maintaining the acceptable performance of breeding as well as non-breeding animals. Neonatal, growing, finishing animals are all required to meet basic livestock nutritional requirements. Animal nutrition is the science of nutrition deals with many a biochemical reactions and physiological processes which convert feeds into body tissues and various activities of living organism. It involves processes such as feed intake, digestion and assimilation or absorption of nutrients into the animal body.

The best way to fulfil this requirement in animals is by creating an optimal nutritional program. The program should focus on removing deficiencies in basal diets.

The deficiencies can be removed through a supplementation program. For example corn- and soybean meal-based diets for swine; milk replacers for calves and lambs; and available forage for ruminants.

An adequate and ideal diet shall incorporate intakes of amino acids (both traditionally classified essential and nonessential), carbohydrates, fatty acids, minerals, and vitamins.

4.4.1. Benefits of Dietary Supplements

Nutrients like arginine, glutamine, zinc, and conjugated linoleic acid are helpful for animal's growth. They not only help in improving fertility and pregnancy outcome but also strengthen the immune system. Neonatal survival and growth, feed efficiency, and meat quality are other benefits of consuming such nutrients.

Dietary supplementation is one of the ways that involves the consumption of such nutrients. It regulates Gene expression and other metabolic issues. Fundamental, as well as applied research, is required to create an optimal program.

A successful nutrition program shall incorporate all nutritionally important minerals. Also, a proper balance of protein, energy, vitamins is required to make it more productive and economical.

- **Reaching Adequate Water Intake:** Water is one of the important things that are crucial to the nutrition program for animals. The livestock may get health issues due to consumption of the substandard quality of water. It is a known fact that the consumption of water is more important than the consumption of food. There must be a good and consistent water supply for running a successful livestock enterprise. The water supply should be of good quality.

Functions of water:

1. Acts as a general lubricant and cleansing agent for different parts of animal body.
2. Provide rigidity and elasticity to body cells.
3. as a solvent for absorption of nutrients & excretion of waste products.
4. Helps in maintaining the osmotic pressure of body.
5. Regulates body temperature.
6. Maintains moistness in body.
7. Transportation of absorbed nutrients to various body parts.
8. It helps in gaseous exchange in the tissues & lungs.
9. The aqueous humor helps to keep up the shape and tension of the eyeball and acts as a refractive medium for light.

The livestock must be supplied enough safe water to suffice there need of the minimum water requirement. This will ensure that the intake of feed is well digested. If there is not enough safe supply of water then the intake of feed in animals will drop and as a result production will fall. This will ultimately result in the loss of money to the producer.

Further, we will be looking for the nutrients that animal need to survive and the balance of different macronutrients like vitamins and minerals in the body.

4.5. WHAT IS NUTRITION?

The process of providing or obtaining the food necessary for health and growth is called Nutrition. In all cultures, races, and ethnicities, there is a common connection called food. When people come together the common thing is the food which is enjoyed by them. Food is considered a way of doing fun. Not just fun, but it is the fuel for our body. Food provides only the energy needed to carry out the daily functions, full energy. Animal body cells take energy from food (Figure 4.1).

Figure 4.1. Animals need high-nutritional food to produce nutritional products.

Source: Image by Bertoli.

There are three macronutrients found in food. These are proteins, carbohydrates, and lipids. A person or an animal needs the correct amount of the above each to be healthy. Water and different micronutrients life vitamins and minerals required by animals too. We are going to learn how much of the cell nutrients find the animals and the importance of each micronutrient.

Livestock provide essential products such as milk, meat, and eggs and, in turn, provide nutrients like iron, zinc, calcium, and Vitamin A. Actually, vitamin B12 can only be found naturally in foods from animals.
Animal-sourced foods are a building block: Good nutrition in the first 1000 days of life is important for proper cognitive as well as physical development throughout life. Researchers at the United Nations suggested animal-sourced foods be included in a child's diet during this crucial period. These are nutrient-dense foods that help to promote good nutrition and prevent deficiencies.

4.5.1. Carbohydrates

Across the globe the food rich in carbohydrates are bread, pasta, rice and potatoes. These are primarily composed of carbohydrates. Basically, carbohydrates are sugars. It is a long chain of sugar that is assembled, for example, starch. In many types of food items, the carbohydrate is removed. But they are very healthy for animals.

Carbohydrate is also essential for humans. Human should get around 45–65% calories from carbohydrates. In the digestive system of humans and animals, carbohydrate is broken down into simple sugar.

The sugar so collected in the system is then used by the body for instant energy that is needed for growth, repair, and generating heat to keep warm. To maintain optimal health, carbohydrates must be consumed daily by the body. Otherwise, the carbohydrate so collected gets stored in the liver as glycogen and eventually gets depleted.

The most important part of the human body is the brain which runs on glucose only, a simple sugar. The reason why people suffer from low blood sugar levels is that it cannot derive energy from any other micronutrients. In simple language, green is nothing without sugar, or it cannot do its function without sugar.

Livestock which is herbivorous; also require carbohydrate in the form of fibre. Fibre is generally found in plant matter. It is in the form of cellulose and is composed of long chains of sugars. All Digestive system of all animals does not break down the fibre largely but is broken by bacteria in the gut.

However, in the case of grazing animals such as cattle, they have extensive compartments in the stomach which is filled with bacteria. Those bacteria help in breaking down the fibre. Fibre helps in keeping the digestive system moving to create faeces. It does so by absorbing water in the body.

Carbohydrate is the organic material containing chemically CHO, which includes starch, sugars, celluloses and gums.

Formed in plants by photosynthesis and used as ready energy source in feed.

Sources of Carbohydrates in animal body:

a) **Crude fibre:** Fibrous, woody, hard material

It contains cellulose, hemicellulose, lignin and pentosans, not easily digested in animal body.

Rumen micro-organisms convert cellulose into simple sugars and fatty acids

Mature fodder, straw and bhusa, dry grass contains the crude fibre.

b) **Nitrogen free extract**:

These includes starch and sugars (soluble carbohydates). These are helpful in making up milk sugar (lactose).

Functions of Carbohydrates:

1. Provide enery more readily than other oraganic constituent
2. Maintains body temparature
3. 1Excess of carbohydates stored as fats, a reserved sorce of energy
4. Important part of blood as a blood sugar.
5. Stored as glycogen in body tissues and liver
6. Induce sweetness to milk
7. Helps in absorption of calcium & phosphorus in growing calves.
8. Helpful in growth & multiplication of micro-organisms in the rumen.

4.5.2. Proteins

Proteins are those macronutrients which is the need of the body for growth and repair. Generally, when one thinks of proteins, the first thing that comes to mind is a protein powder which is taken by gym-goers in the food stores.

But protein is not only for making muscles but also it is an important macronutrient for the body. Depending on the age and the activity level of humans, the requirement of the calories that come from protein is about 10–35%.

Proteins are musically made of amino acids. The body makes amino acids by itself, while some essential amino acids come from the diet one takes. There is a different composition of amino acids in protein sources found in different types of food.

There is a complete set of amino acids which is usually found in meat. But in vegetarians, all the essential amino acids are absorbed by a different variety of food items. This is the reason that it is difficult for herbivorous livestock to get all the essential amino acids as they only eat plants. So, they are needed to be given feed supplements that carry the entire social message for attaining proper growth and development.

Functions of proteins

1. Protein makes up 17 to 20 % of the animal body and 3.6 % of the milk.
2. Makes up new tissues and muscles in the body
3. Repairs the loss of body tissues of muscles.

4. Helps in growth and development of body.
4. Helps in production of enzymes & hormones.
6. Helps in production of wool, egg and meat.
7. Excess protein provides energy.

4.5.3. Lipids

Lipid is another word used for fats. Basically, lipids are fats present in different food items such as oil, butter, and various other goods produced using them. In terms of health and nutrition, lipids are not considered to be good; however, they are really critical for the human body.

Lipids are mainly responsible for the long term storage of energy in the body, cushion different organs, and also insulating brain cells to aid them to send message to the body in an efficient way (Figure 4.2).

There are two types of lipids:

- Saturated lipids; and
- Unsaturated lipids.

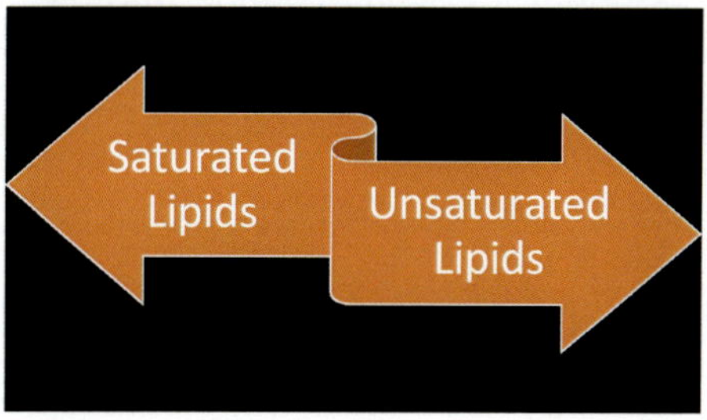

Figure 4.2. Types of lipids.

Saturated lipids are mainly derived from animal fats and can be harmful for the human body. On the other hand, unsaturated lipids are essential for good health and thus should be included in a healthy diet. These lipids come from plant-based fats like avocado and oil. Nevertheless, the amount of fat present in an animal's diet must be balanced with various other macronutrients. Human diets must no more than 20%–25% of calories derived from lipids.

Classification of lipids:

Lipids are divided into 1.Simple lipids, 2. Compound or conjugated lipids & Derived lipids.

1. Simple lipids are fats/waxes(Esters of fatty acids with glycerol).
2. Compound lipids are glycolipids and phospholipids.

Derived lipids are essential oil, steroids and sterol.

- Sources of fats:
- Oil seeds- cotton seed, linseed, groundnut, mustard, til, cereals etc.

Functions of Fats:

1. Provides energy 2.25 times more than carbohydrates.
2. Reserve source of energy.
3. Helps in formation of milk fat.
4. Makes up approx. 20% of animal body.
5. Essential constituent of protoplasm.
6. Helps in absorption of carotene.
7. Helps in availability of vitamins A, D, E, K.
8. Provides essential fatty acids like linoleic acid, linolinic and arachidonic acids.
9. Acts as shock absorber & prevent injury to internal organs.

4.5.4. Water

Water is known as the most essential ingredient for life. It is very significant that scientists looking for life on some other planets first search for the signs of water. Every human knows that no life can exist without water. The question arises, how much water does a human need, as water intake largely depends on age, activity level, and environmental conditions.

In terms of livestock, cattle have high water requirements; cows require up to 90 liters of water per day, especially during hot summer days. Comparatively, humans are advised to drink around three liters of water. Though, hot summer days and a great level of physical activity results in more water loss and thus requires a greater water intake.

Sources of water for animal body:

1. By voluntary intake: through drinking water
2. By feed: Through green fodder
3. Metabolic water: Water produced due to oxidation of nutrients in body (5-10%)

Functions of water:

1. Acts as a general lubricant and cleansing agent for different parts of animal body.
2. Provide rigidity and elasticity to body cells.
3. As a solvent for absorption of nutrients & excretion of waste products.
4. Helps in maintaining the osmotic pressure of body.
5. Regulates body temperature.
6. Maintains moistness in body.
7. Transportation of absorbed nutrients to various body parts.
8. It helps in gaseous exchange in the tissues & lungs.
9. The aqueous humor helps to keep up the shape and tension of the eyeball and acts as a refractive medium for light.

Good health requires proper nutrition, which comes from a balanced diet. The animals are fed mostly with the processed food. Processed food has become today's staple diet for animals due to its long shelf life and easy availability.

Processed foods lack many essential nutrients since they are cooked, refined, and in fact denatured. It is appropriate to call them Industrial Byproducts instead of Processed Foods. Various Preservatives, dyes, artificial flavours, sweeteners, colours, synthetic vitamins, sugars, and hydrogenated oils are mixed as chemical concoctions.

Not only these but some processed foods are prepared from genetically modified (GM) ingredients or are made mostly from adulterated ingredients.

Processed foods are cooked at a very high temperature. Cooking food at high temperatures destroys enzymes, fatty acids, amino acids, vitamins, and minerals. Generally, this is not recommended for an animal's diet.

Unfortunately, they are fed the same food which lacks all essential nutrients daily. Not only this, the same meal is fed to our pets of the same brand every day, which is not only harmful but also mentally tiring. Imagine humans being fed the same meal during their entire life, no one would be happy and healthy.

According to Integrative Veterinarian, Dr. Karen Becker of Mercola Healthy Pets, "The pet food industry is rapidly becoming a big business. Big business to the tune of $15 billion worldwide. What most pet owners don't realize is that the pet food industry is actually an extension of the food and agricultural industries."Pet food provides these giant multinational corporations with a very convenient and very profitable 'built-in' market for wastes leftover from their human food production. But here's the problem… your beloved pet was never designed to eat 'wastes' (Figure 4.3).

Figure 4.3. Water: Essential intake for animals.

Source: *Image by Weld Laboratories.*

Pet food manufacturers may use many ingredients as their selling point based on the fact that they are full of nutrition. One of the best examples would be chicken. Many Pet Food bags list Chicken as their main ingredient. It sounds appealing and healthy, but the reality is different. The chicken used in Petfoods is not only unhealthy due to it being processed but also because of the quality of chicken used.

It is also not stated as which part of chicken is used in pet food. Meat meals, by-product meals, or meat-and-bone meals are common in pet food. Also, these parts could be rendered and not fresh, which is again very harmful to animal's health.

In animal food, generally, the ingredients are boiled in a large vat for several hours at a very high temperature. It is stated that many valuable nutrients like vitamins, enzymes, coenzymes, phytonutrients, and proteins are inactivated or even destroyed at temperatures above 118°C. Whereas, it has been observed that processed foods are cooked at high temperatures which may reach up to 500°F.

The way the food for animals is cooked and the type of ingredients used in the process is very harmful and may cause various diseases in animals. Diseases like cancer, arthritis, allergies, inflammation, and autoimmune diseases are very common.

As per Dr. Becker, many diseases can be there due to denatured proteins. It leads to food allergies and intolerances and, also inflammatory bowel disease.

Every living creature has the right to be healthy. Just like Human beings, animals also suffer health issues on the consumption of a consistent diet of processed foods. This is because of the destruction of important nutrients through food processing.

Healthy cell division is required for a healthy body, to be healthy, it is very important to include living foods for the consumption. Living food means raw food or we can say unprocessed food. Incorporating a raw diet is very important for humans as well as animals.

There are many advantages of consuming a raw food diet, but unfortunately, a lot of the population does not have the time, know-how, or inclination to create their own raw meals.

Living creatures being fed wholesome foods are healthier than a creature that is are nutritionally depleted. Raw foods are easily available and accessible, giving the options to live a healthy life. Nowadays raw foods are available in frozen form too.

However, storing raw food for longer than a few months is not advisable. When kept frozen for a longer time, nutrients may degrade with time. It is important to maintain optimum nutritional value, including enzymatic activity in raw foods; hence storing them for a longer period is not a healthy option.

To address health issues of the kidneys, liver, heart, and endocrine system, various animal food manufacturers along with veterinarians are researching on creating optimal food supplements for animals. Darwin's Natural Pet Products is working with veterinarians to design a prescription line that will include specific formulas to address the health problems.

Another option for pet owners other than buying food from the market, they can prepare their pet's food on their own. Books like Dr. Becker's Real Food for Healthy Dogs and Cats provide various recipes for pet food.

Other than a balanced diet including raw food and eliminating processed food, there are other options too in order to provide nutrition to animals.

One of them is supplementation. Supplements are a good source of many essential nutrients.

Just like human beings, consuming supplements are equally beneficial for pets; it helps to meet an animal's nutritional demands. However, due care needs to be given when choosing supplements. The quality of the supplement is equally important like a balanced diet.

Whole foods, botanicals, organ, and tissue extracts (including photomorphogenesis, which are the blueprints of a cell) are usually the main ingredients of quality supplements. These whole food concentrates are created from a low-temperature, high vacuum process.

This is done to preserve the ingredients' nutrients, including enzymes. Enzymes are responsible for all the metabolic reactions which help from the digestion of food to the synthesis of deoxyribonucleic acid (DNA). Standard Process, Inc., is one of the producers of such supplements for animals.

Standard process whole food supplement addresses health at the cellular level for all their products. For a cell to function properly, specific quality and quantities of the nutrients should be maintained. Without specific nutrients in specific quality and quantities, cell function may be impaired.

Food supplements for animals nowadays lack these basic requirements, which can lead to physical symptoms if these requirements go unaddressed for a longer period. Such unwanted symptoms may be stopped by medication. However, a permanent solution is the need of the hour which must be analyzed deeply and nutritionally.

Vitamin intake is one of the important parts of maintaining animals' health. While choosing the kind and type of vitamin supplement to be used, it should be kept in mind that the quality of the vitamin should be good, and it should be taken in proper quantity. More quantity does not necessarily mean more nutrients. In the case of vitamins, consuming more quantity of it may lead to some serious diseases.

Another point to be noted while consuming vitamins is the way it is being consumed. Whenever vitamin is consumed in its whole food form, it has shown far more effective result than a large dose of isolation or a fraction of the whole vitamin.

Animal bodies can be severely harmed when the vitamins are consumed in large dose instead of small fractions or whole food forms. Standard Process concentrates foods for use in supplements, which supplies a multitude of components that work synergistically to provide optimal nutrition for good health. Glandular extracts are useful when it comes to providing essential nutrients. Standard Process whole food supplements provide glandular therapy with the use of glandular extracts. This therapy helps in healing animal's glandular tissues that have compromised function. It not only recovers and heals an animal's tissues but also helps in returning to its normal function.

The building blocks that are critical to maintenance, regeneration, and repair are often recognized when the nutritional requirements of a body are addressed carefully, whether it's for an animal or a human.

4.6. THE IMPORTANCE OF ANIMAL NUTRITION IN LIVESTOCK PRODUCTION

Just like human beings, animals also need proper nutrition to grow, develop, and reproduce properly. Also, it requires strong immunity to fight off infections in animals. Hence, a balanced diet containing all the necessary nutrients, fluids, minerals, and vitamins shall be provided to livestock animals just like human beings.

Proper nutrition gives animals the vigour to not only grow but also reproduce hence all these advantages ultimately lead to more profitable and sustainable agriculture.

Every animal's dietary requirements are different at different stages of life, hence while choosing the diet for an animal, it should be made sure that all the essential dietary requirements are provided that are required at different stages of an animals' life.

For example, a lactating cow's nutrients requirement would be different from the requirements of a calf. Similarly, the nutrients essential for a pig may not be essential for cattle, you can always boost your feeds' specific nutritional content by adding feed supplements for livestock. Let's take an in-depth look at the importance of animal nutrition in livestock production.

4.6.1. Prevent Malnutrition, Deficiencies, and Diseases

Livestock animals are affected by hundreds of illnesses that are caused by a lack of nutrition. These are also due to some lack of vitamins, malnutrition, and minerals. The growth and development in the animal are severely affected by deficiencies and malnutrition.

This is the reason that these animals produce less. Or in other words, the production is of low quality or quantity due to malnutrition. There are instances that these may cause disorders, fatalities, and irreversible health conditions.

Most of the animals nowadays are given feed on concentrates that are specially designed for maximizing yield. Due to this, most of the animals lack nutrition, and hence they are prone to diseases that are related to nutrition. Concentrates may lack some essential vitamins and minerals. That is the reason these diseases are a problem due to these modern farming techniques.

To maintain good health in livestock, they must be given high-value supplements such as B12 supplements. They can also be provided with salt-based additives. These salt-based additives contain various micro and macro minerals. For example, in order to increase lactation and growth in cattle, it is essential to meet the high energy demand by including these supplements.

4.6.2. Improve Breeding

Animal's reproductive health is affected by numerous factors. One of the most important factors is the nutritional value they get from their food. Other factors may be listed as the feeding patterns, supplementation, the quality, and quantity of rations, and so on.

This has been a topic of research for years now, and in fact, animal's experts have published their studies depicting the direct correlation between nutrition and reproductive performance in animals.

To maintain health and improve fertility, minerals like calcium, zinc, magnesium, selenium, and manganese play a very important role. Deficiencies of these minerals may directly affect fertility in animals. It has been observed in cows that deficiencies of such minerals have reduced the fertility rate by increasing the risks of placental retention and mastitis and upsetting the balance of gestation and parturition hormones (Figure 4.4).

Figure 4.4. Breeding improvement is highly desired.

Source: *Image by The Humane Society of the United States.*

In various cases, not only fertility but improper nutrition also leads to poor fetal development or stunted growth after birth. It has also increased calf mortality rates in severe cases. It also affects the reproductive vitality of male animals. Animals like Bulls, stallions, roosters require a special diet as well as proper supplementation to ensure their spawns health and viability.

4.6.3. Boost Yields

Higher productivity and good health are the main advantages of feeding proper nutrition to animals. These ultimately lead to high yields not only in terms of quality but also quantity. Well-fed healthy cattle and poultry are likely to produce more eggs, milk, and meat of good quality than a malnutrition animal.

With time, consumer's preferences are also changing not only in terms of the quality of food product but also the way the food products are being produced. The demands for sustainably farmed produce have increased drastically.

Not only in Farming but also the livestock industry is expected to follow humane farming practices. In such types of farming practices, the focus remains on improving the production of animals, hence, it is stated that no animals should be made to consume harmful substances.

Many producers use unscrupulous means to boost their farm produce in order to provide more yields in the market to sustain competition. However, it is believed that instead of using shortcuts for higher yield, animals should be fed the right way in the right quality and quantity, and this will surely lead to quality products with high production. As rightly said, a rich diet not only improves yields but also enhances the quality of livestock production.

4.6.4. Bottom Line

A Farm's success is based on the feeding patterns and the care their animals get. Just like Human Body, an animal's body is also very sensitive to what they consume and how they consume it. Hence the importance of proper nutrition for livestock production cannot be refused on any ground.

Food not only determines livestock's overall health but also ensures higher productivity and yields. The only way to ensure farms growth and continued success is to provide animals with all the essential nutrition in the right quantity and quality and at the right time.

Animal husbandry practices evolved along with the evolution of the human population. Just the way the human population grew from 4 million people 10,000 years before the present (YBP) to the 5.85 billion people that occupy the world today, animal agriculture too evolved.

It all started with the domestication of few mammals that were initially available, and continuous studies and research lead to a whole developed sector of Animal Husbandry where it is one of the primary sectors of income for many in the whole world.

Despite various modern techniques and technological advances, traditional methods of Animal husbandry are followed in most parts of the world. There are only a few domesticated species available.

Technological advancement in terms of physiological manipulations, nutritional provisions, genetics are developing with time, and it is expected to see some changes in animal agriculture practices.

In 1996, the world population of domesticated grazing livestock was approximately 3.3 billion, an increase of 1.1 billion during the last 5 decades despite a relatively constant total amount of grazeable land.

The bulk of this increase during the last half of the 20th century is due to a near doubling of cattle, sheep, and goats on the continents of Asia and Africa. The impacts of unmanaged grazing animals have been substantial through the course of human history.

The modem principles for proper management of grazing lands developed during this century in response to destructive impacts of unmanaged grazing throughout the world.

4.7. GRAZING ANIMAL IMPACTS

Grazing is the foraging on plant materials. Those plant materials are the characteristic feature of livestock use of pastures and rangelands. Grazing impacts plant's growth and development in many ways. There are the direct as well as indirect aftereffects of grazing.

Some of direct impacts would include consumption of plant issues, trampling of soil surfaces, and disruption of cryptogamic organisms. Not only this but removal and excretion of nutrients by animals are also considered as one of the factors directly impacted by grazing.

Alterations of numerous plant morphological and physiological attributes are very common and are indirectly affected by grazing. Indirect effects include spatial and temporal redistribution of soil nutrients. It has also been observed that grazing indirectly causes alterations of plant gene frequencies, changes to structural features of plant communities, and alterations of water and energy (Figure 4.5).

Figure 4.5. Grazing serves a key role in livestock production.

Source: *Image by Hobby Farms.*

The environment is dynamic, and thus the variations in the impacts of grazing can vary tremendously. It could be seen due to changing environmental conditions. The Concurrence of environmental stresses like fire and drought compounds the effects of grazing.

Predicting and analyzing these effects in specific environments might not be useful due to the dynamic nature of the environment. It is often episodic and reflects the occurrence of mitigating site conditions.

Herbivory is a natural process that can have complex impacts in natural settings (Collins et al., 1998); however, unmanaged grazing by livestock can have extreme and negative effects on ecosystems at local, regional, and continental scales (McNaughton, 1993).

Most of the continents are overgrazed by the livestock, due to which there is damage to the grazing land. This damage to the land due to overgrazing was seen in a relatively short period, just within one or two decades. The nutrients in arid condition of the land is poor and becuause of this the lands are not getting recovered quickly. The land is instead becoming nonexistent (Schlesinger et al., 1990).

The environment having high erodible soil or, in other words, low primary productivity is not considered suitable for grazing by domestic animals. However, with some consistent efforts of humans, the managed grazing system has improved the grazing level by the animals (Heitschmidt and Stuth, 1991).

4.7.1. Managed Grazing

Managed grazing is behaviour of grazing by the animals that are organized in such a way that is under some control of the management by humans. This management control is often at a pasture scale. It can vary from less than one hectare in humid climates to over 10,000 hectares in arid climates.

The timings, frequency, intensity, and duration of grazing are the key factors that determine the management control for the managed grazing (Trlica and Rittenhouse, 1993). The functions and the structure of the grazing land ecosystem are affected by each of these factors.

The most common factor that controls the grazing land ecosystem is the intensity of its use. The impact of other factors such as season of use and spatial distribution of grazing animals are also recognized well. There is a well-articulated management principle (Holechek et al., 1997).

The grazing system is a management program in which the key factors of grazing are controlled to some degree. There are four main categories in which the grazing system falls: true nomadism, semi-sedentary, transhumant, and sedentary (Williams, 1981).

Each category is different from the other, but the system of grazing is characterized by some level of management over the other. It is important to know that the evolution of the grazing system was there during the past 10.000 years. Also, the plant vertebrate interaction has occurred over millions of years.

The grazing system has been developed to minimize the effects of livestock grazing or to bring the historical grazing behaviour of the native herbivores back. There is no combination of control factors that has the advantage of using over the other factor of grazing control. There is no evidence that any factor has the power to compensate for the mismanagement or improve the condition of the resource.

A particular system has the benefit over the other when there is human involvement in the management of the land, which helps in deployment. Then the benefits that are secondary to it are the improved animal distribution associated with watering tanks, the construction of new fences in order to conserve the sensitive areas. This comes under the implementation of the system.

Rather than bringing back the existing state, today's management of the grazing system is making efforts towards achieving desired resource conditions (Westoby et al., 1989). In both developed and a less developed nation, this change is taking place.

In the developed nation, where there is more emphasis on the use of grazing land and in a less developed nation, where livestock production is the main factor in subsistence agriculture.

The main issue is that whether the animals that are grazing can be used for landscape management along with the agriculture products. There can be numerous objectives of the management that has a variety of goals such as decreased use of harvested forages, watershed functions being improved, and the biological control of introduced plant species.

There is a need for good management of these objectives of grazing by livestock for minimizing the negative effects of grazing.

4.8. CONCLUSION

Animal husbandry has a long history that started from the time of the Neolithic revolution. This was the time when animals were first domesticated. Domestication was not a single phenomenon but a series of process that was repeated after time periods.

Across cultures, food is a common connection. Among all races and ethnicities, only food is the one that is the only feeding source to all the types of domesticated animals. It is fuel for an animal's body. Along with this, nutritional requirements also matter a lot to produce good results.

Animals that were domesticated were given nutritional feed to get good products out of them. Regularly feeding them good food is the basic requirement.

REFERENCES

1. Animalscience.tamu.edu. (2021). *Animal Nutrition| Animal Science*. [Online] Available at: https://animalscience.tamu.edu/academics/nutrition/ (accessed on 15 March 2021).
2. Clutton-Brock, J. (1981).Domesticated Animals from Early Times. University of Texas Press, Austin, Texas. A general text summarizing the history of animal domestication from cats to cattle.
3. Collins, S.L., Knapp, A.L., Briggs, J.M., Blair, J.M., & Steinauer, E.M.(1998). "Modulation of Diversity of Grazing and Mowing in Native Tallgrass Prairie." *Science*, 280: 745-747.
4. Department of Animal Science, (1959). *History of Animal Husbandry Department.* [Online] Available at: https://www.ans.iastate.edu/about/history/animalhusbandry (accessed on 15 March 2021).
5. Dregne, H., Kassas, M., and Rozanov, B.(1991). "A New Assessment of the World Status of Desertification." *Desertification Control Bulletin*, 20: 6- 18.
6. FAO. (1996). Production Yearbook. v50. Food and Agriculture Organization of the United Nations, Rome, 1996. Compilation of agricultural statistics.
7. Havstad, K., *Animal Husbandry*. [eBook] Available at: https://jornada.nmsu.edu/bibliography/02-107.pdf (accessed on 15 March 2021).
8. Healthycellsmagazine.com. (2013). *The Importance of Proper Animal Nutrition*. [Online] Available at: http://www.healthycellsmagazine.com/articles/the-importance-of-proper-animal-nutrition (accessed on 15 March 2021).
9. Heitschmidt, R.K., and Stuth, J. W., (1991). Grazing Management: an Ecological Perspective. Timber Press, Portland, Oregon.
10. Holechek, J.L., Pieper, R.D. & Herbel, C.H. (1998). Range Management Principles and Practices. 3rd ed. Prentice-Hall, Englewood Cliffs, NJ 07632, USA.Pp 542.
11. McEvedy, C., & Jones, R., (1978). *Atlas of World Population History*. Penguin Books Ltd.; New York, NY.
12. McNaughton, S.J. (1993). "Grasses and Grazers, Science and Management." *Ecological Applications*, 3: 17-20.
13. Patil, P.V., & Patil, M.K., (2021). *Milk production management*. Jaya publishing house, Delhi, India.

14. ProviCo. (n.d). *The Importance of Animal Nutrition in Livestock Production-ProviCo Rural.* [Online] Available at: https://www.provicorural.com.au/the-importance-of-animal-nutrition-in-livestock-production/#:~:text=Like%20humans%2C%20livestock%20animals%20need,immunity%20to%20fight%20off%20infections (accessed on 15 March 2021).
15. Schlesinger, W.H., Reynolds, J.F., Cunningham, G.L., Huenneke, L.F., Jarrell, W.M., Virginia, R.A., and Whitford, W.G. (1990). "Biological Feedbacks in Global Desertification." *Science*, 247: 1043- 16 1048.
16. Study.com. (n.d). *Animal Nutrition: Requirements and Importance.* [Online] Available at: https://study.com/academy/lesson/animal-nutrition-requirements-importance.html (accessed on 15 March 2021).
17. Trlica, M. J., and Rittenhouse, L.R. (1993). "Grazing and Plant Performance." *Ecological Applications*, 3: 21-23.
18. Vigne, J., (n.d). *The Origins of Animal Domestication and Husbandry: A Major Change in the History of Humanity and the Biosphere.* [Online] ScienceDirect. Available at: https://www.sciencedirect.com/science/article/pii/S1631069110002982#sec0010 (accessed on 15 March 2021).
19. Vila, C., Savolainen, P., Maldonado, J.E., Amorin, I.R., Rice, J.E., Honeycutt, R.L., Crandall, K. A., Lundeberg, J., & Wayne, R.K. (1997). "Multiple and Ancient Origins of the Domestic Dog." Science, 276: 1687-1692.
20. Westoby, M., Walker, B., and Noy-Meir, I. (1989). "Opportunistic Management for Rangelands not at Equilibrium." *Journal of Ranae Management*, 42: 266-274.
21. Williams, O.B. (1981). "Evolution of Grazing Systems." In Grazing Animals, edited by F.H.W. Morley. Elsevier Scientific Publishing Co., Amsterdam, pp. 1 - 12.

Chapter 5

Feeding Standards for Animals

CONTENTS

It is important to note that feeding standards are most important to ensure the growth and development of animals. In the chapter, feeding standard for animals, several types of feeding standard are discussed. It also highlighted the use or practical application of feeding standard. It also shed some light on the nutrient requirement for maintenance. It also demonstrates the importance of minerals and vitamins for the growth and maintenance of animals. It also explains in detail the feeding standard of buffalos and cows. In the end, it discusses some of the usefulness and limitations of feeding standard.

5.1. INTRODUCTION

Feeding standards can be described as the statements or the quantitative descriptions of the exact amount of one or more nutrients that are required by the animals. In an elaborative manner, feeding standards are defined as the tables that show the quantities of nutrients that are needed to be fed to different classes of livestock for various physiological functions such as growth, lactation, maintenance, wool growth, egg production, etc.

In general, the nutrient requirements are expressed in terms of quantities of nutrients needed per day or as a diet percentage. While for dairy animals, nutrient requirements are expressed as separate body functions. On the other hand, for poultry and pigs, it is expressed as combined requirements of maintenance along with certain other body functions.

It is generally seen that feed is the single most significant element of animal production regardless of species and production system in the context of economic consideration as it accounts for up to 70% of the entire cost.

Ensuring the availability of desired type and quantity of livestock feed in a sustained manner highly influences the overall production system of livestock. Optimum usage of available feed lessens feeding costs and intensifies the economic viability of the livestock process (Makkar and Beever, 2013).

It is expected that in the future, the global demand for livestock products will likely increase and is believed to surge by 60% by the year 2060. By 2021 itself, developing countries will likely produce almost half of the milk of the world with India dominating the output.

Hence, it is right to assume that the livestock sector is under immense pressure to really enhance the capacity of nutritious and safe feed, make them accessible to the animals in an adequate form to provide and produce a comparatively higher amount of good quality animal products. But, ensuring higher production would not be enough; its compatibility with animal product quality and safety, animal health and welfare, environment friendly and biodiversity are also gradually demanded. So, in the present era, an animal diet may be expressed as the one that is free from poisonous components, balanced in nutrients, meets the production goal, produces animal products that will be safe for human consumption, along with integrating principles of sustainability (Makkar et al., 2014). Therefore, keeping an eye on the quality of animal feed would do the majority of the tasks in accomplishing these objectives. Feed quality can be defined as "any of the features in the feed that makes it what it is" and "the degree of excellence which the feed possesses." A good quality feed can be defined as the one that will supply all the essential nutrients in passable quantity with high digestibility and palatability. The quality control objective of feeds is to make sure that a consumer gets access to feeds that are true to their nature, unadulterated, include less processing, provide the intended nutrients, and produce wanted results in the production of animal. To achieve this control in quality, the feeds have to necessarily undergo various steps of quality evaluation. Feed evaluation is basically the testing of feed quality which is responsible for conveying information and data on the composition of feed or feed ingredients as well as their suitability for poultry and livestock. Feed quality can be ascertained by knowing the physical nature, characteristics, and the actual composition of the feed both in terms of anti-nutritional and nutritional constituents. It is generally seen that feed is made up of numerous ingredients, which are largely clustered into providers of energy (oils, fats, and carbohydrates), vitamins, protein (amino acids), minerals, and products for overall quality enhancement.

Feeding standards of the world for feeding different categories of livestock:

Name of the Country	Protein	Energy
NRC(USA)	CP, DCP	TDN, DE, NE
ARC(UK)	DCP, AP	DE, ME
SCANDINAVIA	DTP	FEED UNIT
GERMAN	DCP	SE
INDIA	DCP	TDN, ME

Where; CP is the crude protein; DCP is the digestible crude protein; DTP is the digestible true protein; AP is the available protein; TDN is the total digestible nutrient; SE is the starch equivalent; DE is the digestible energy; ME is the metabolizable energy; NE is the net energy.

Feed ingredients like cereal grains such as wheat, maize, sorghum provide energy; while oilcake and meals of soybean, mustard, groundnut, linseed, sunflower, and sesame and animal byproducts such as meat meal, fishmeal, provide protein.

Byproducts such as rice bran, wheat bran, rice polish, maize gluten, oiled rice bran, broken gram, gram husk, etc., are also used in animal feeds to offer the needed nutrients.

In a way to recognize what proportion of these ingredients should be comprised in the diet during feed formulation, the first task is to evaluate the ingredients, to see what nutrients they embrace in what quantities.

After the preparation of diet, it may also be essential to appraise the whole product, to evaluate its suitability for the class of poultry and livestock to which it will be fed.

It may also be essential to initiate random checks on the quality of stored finished feeds for any decline in nutritional quality. Feed evaluation is one of the most important and critical tasks that offer distinctive types of information, as required by farmers, nutritionists, and traders.

The various feeding standards that are prevalent today prescribe the amount of dry matter (DM) and the proportions and amounts of the several common digestible nutrients—carbohydrates, protein, and fat—which, it is widely believed, are required for the highest physiological well-being of the several classes of animals.

The feeding standards may vary on the number of factors such as the local environment, routine activities of animals and body type. So, it can be said that presently, no feeding standard can give any concrete advice.

For instance, we know that an acceptable supply of the several essential mineral elements is crucial for life, and that these elements should be supplied more copiously to animals that are pregnant, growing, or giving milk than to mature animals being fattened or maintained. Having a proper comprehensive plan for each farm type and location will work best in this case.

5.2. TYPES OF FEEDING STANDARD

In the feeding standards, there are basically two terms that have been using for a long. One of them is the nutrient allowance while another is the nutrients requirements. The former one gives an additional allowance of nutrient over the requirement, which gives a slight margin of safety. On the other hand, the latter one gives the requirement for optimum production (Figure 5.1).

Figure 5.1. Sustainability in animal feeding.

Source: *Image by Agriskills.*

Feeding Standards:

1. Comparative type:
   i. Hay standard;
   ii. Scandinavian "feed unit" standard.
2. Digestible-nutrient system:
   i. Grouven's feeding system;
   ii. Wolff's feeding standard;
   iii. Wolff's Lehmann feeding standard;
   iv. Haeckers's feeding standard;
   v. Savage feeding standard;
   vi. Morrison standard;

vii. National Research Council Standard;

viii. Japanese feeding standard;

ix. Indian standard.

3. Production-value type:

i. Kellner-feeding standard;

ii. Armsby feeding standard;

iii. Agricultural Research Council Standard.

5.2.1. Comparative Type

1. **Hay Standard:** In the year 1810, a German scientist Thaer recommended that different feeds must be compared by using 'meadow hay' as a unit. This standard provided that, 100 lbs of meadow hay was equal to 91 lbs of clover hay or 200 lbs of potatoes, 625 lbs of mangles (in terms of nutritive value).

He compared different feeds with Medow hay as a unit. In this standard the nutritive value of feed and physiological requirements of animal were not considered. It was completely based on practical feeding experience. This is also called as hay equivalent feeding standard

2. **Scandinavian "Feed Unit" Standard:** In the year 1884, the Scandinavian feeding standard was formulated by professor Fjord. In this standard, only one factor 'the feed unit' was considered. The value of one pound of common grain, for example, corn, wheat, or barley, was given as one unit value while the value of all other foods was based upon this system.

As per this standard, one feed unit is needed for every 150 lbs of body weight and an extra unit for every three pounds of milk production. However, as the grains in different countries are of different types, the feed unit must also be different. Therefore, the Scandinavian units are not effectively applicable in every country unless experiments are conducted with their own grains.

5.2.2. Digestible Nutrient System

1. **Grouven's Feeding Standard:** Grouven, a German chemist published his feeding standard with crude protein, fat, and carbohydrates contained in the feed as the basis of the standard in the year 1859. Based on the fundamentals of this standard, a cow

weighing 1,000 lbs must be fed with 28.7 lbs of DM comprising 2.67 lbs of crude protein, 14.55 of crude carbohydrates and 0.6 lb of crude fat.

Sooner the Grouven, Henneberg, and Stohaan standards found that the total nutrient present in a feed did not from an accurate or precise guide to its value. With different feed, the proportion of digestible parts varied and as a result, the digestible nutrient would be more valuable. So, because of this defect, Grouven's feeding standard is now abandoned.

2. **Wolff's Feeding Standard:** Dr. Emil Von Wolf in the year 1864, proposed a standard based on digestible fats, digestible fats and digestible carbohydrates present in feedstuff. His standard for dairy cows weighing 1,000 lbs. was 24.5 lbs. of DM comprising 2.5 lbs. of digestible carbohydrates and 0.4 lb. of digestible fats. This standard proposes a nutritive ratio of 1: 5.4. Through an improvement over the standard of Grouven, this standard yet does not take the quantity and quality of milk produced into account.
3. **Wolff's Lehmann Feeding Standard:** In the year 1896, Dr. V. Lehmann of Berlin modified Wolff's standard. Till then Wolff's standard was in great use. He additionally considered the quantity of milk produced, but somehow failed to consider the quality of milk.
4. **Haecker's Feeding Standard:** For the first time, an American worker, Haecker considered the quality as well as quantity of milk produced while formulation a milk standard. Also, he was the first to separate the requirements of maintenance, reproduction from the requirements of production.

His standard comprised digestible crude fats, carbohydrates, and proteins. After some time, it was expressed in terms of digestible crude protein and total digestible nutrients (TDN).

5. **Savage Feeding Standard:** Savage, another American scientist concluded that the standard given by Haecker was too low, mainly in protein. He further recommended that, in the case of milking cows, at least 24 lbs. of DM should be given for an average cow.

It is suggested that the nutritive ratio should not be wider than 1:6 or narrow than 1:4.5. Around two-thirds of the DM must be from the roughages while one-third from the concentrates. As a result, the requirement of protein was increased around 20% above the standard of Haecker. In this feeding standard the quantity and quality of milk produced as well as the different milk fat levels were taken into consideration.

6. **Morrison Feeding Standard:** Morrison F. B. observed that stockmen are spending huge sums of money for completely unnecessary amounts on the protein supplement, and therefore reducing their profits considerably. As a result, he endeavoured to combine in one set of standards what seem in the judgment to be the best guide available in the computation of rations for the various classes of stock.

These standards were first presented in the 15th edition of "Feeds and Feedings" published in the year 1915 and were then called "Modified Wolff and Lehmann standard." Soon, they came to be known as the "Morrison Feeding Standard." These standards have expressed in terms of DM, Digestible Protein (DP) and TDN. This standard was further modified and the requirements for Ca, Phosphorus and carotene were included.

7. **National Research Council (N.R.C.) Standard:** The standard consists of DP as well as TDN along with recommended requirements for phosphorus, calcium, vitamin D, and carotene for beef cattle, dairy cattle, poultry, pigs, horses, sheep, dogs, laboratory animals, and so on. It is supposed that these N.R.C. reports showing in each case are the pooled judgment of a group of experts in the field of species in question.

At present, in many countries N.R.C. standards are followed, where they use DE for swine and horses, ME for poultry, DE, ME, and TDN for sheep, ME, TDN, and NE_m and NE_g for beef cattle and dairy cattle, values of DE, ME, TDN, NE_m, and NEg for growing animals with additional values as NE_l for lactating cows. From time to time, the NRC revises these feeding standards in keeping with new information and changing feeding practices.

8. **Japanese Feeding Standards for Dairy Cattle:** Maintenance requirements based on live weight rose to the 0.75 power of maintenance that is 37.37 g TDN/kg or 116.3 kcal ME/kg (equivalent to 0.58 lb digestible Protein. 8.1. Lb TDN of 11.4 Mcal ME per 1,000 lb cow). Nutrient requirements for milk production were calculated based on 154 parts digestible Protein per 100 parts milk protein and 1,444 kcal ME per 1,000 kcal milk energy.

9. **Indian feeding standard**: This standard is given by ICAR. It includes the requirement for cattles in terms of DCP, TDN, NE, and also for Calcium, Phosphorus and Carotene. Whereas, separate standards have been given for Poultry, Sheep, Pig and

goat. The nutrient requirement for Poultry is expressed in terms of Crude Protein and ME. The Indian feeding standards for cattle are based on Morison feeding standard. A separate standard have been proposed by ISI. Which is also called as BIS (Buro of Indian Standards) for poultry.

5.2.3. Production Value Type

1. **Kellner Feeding Standard:** Kellner a German scientist, in the year 1907, investigated a feeding standard based upon "starch" as the unit of measurement. He not only just considered the digestibility of the feeds as calculated from the amount lost in urine and feces but also the complete loss from the body comprising energy expended in chewing and digestion. For measuring the amount of energy lost from the body as heat, Kellner devised a respiration apparatus.

Here, heat is indirectly determined by finding the precise amount of carbon dioxide (CO_2) gas evolved or by calculating the amount of oxygen gas used up in oxidation which occurs inside the body. Animals breathe through an airtight mask placed over its nose and mouth. As per this system, 1,000 lbs. Animal requires 0.6 lb. of DP and 6.35 lbs. of starch equivalent.

In turn, this starch equivalent can be converted into energy by a method worked out by Armsby and Kellner. For any of the feeds, the composition of which known be converted to starch equivalent by using the following factors:

i. Dig Protein X 0.94 = S.E.

ii. Dig. Carbohydrates and fiber X 1.0 = S.E.

iii. Fat from coarse fodder X 2.1 = S.E.

iv. Fat from cereal grain X 2.1 = S.E.

v. Fat from oil seeds X 2.4 = S.E.

The starch equivalent is defined as the amount of pure starch required to produce the same amount of fat which is produced by 100 kg of feed.

2. **Armsby Feeding Standard:** Armsby standard is mainly based on true protein and the net energy (NE) values. By using respiration calorimeter, Armsby determined the total energy requirement for digestion, mastication, assimilation along with the quantity of gases and heat released through the excretory channels.

Hence, after considering different losses of energy like gases, urine, feces, and in the work of digestion, Armsby was able to estimate the amount of total energy available for productive means. He expresses his standard in two different factors that is true protein and terms of NE.

However, a very common limitation of the Armsby standard is that the cost of determining the requirements of the animals and the total energy in different feeds is quite high. The total energy values of few feeds had been determined, and most of them have computed from the table of Morrison's digestible nutrients. Thus, Armsby standard is not as widely used as are the standards based on digestible nutrients.

3. **Agricultural Research Council (A.R.C.) Standard:** In UK, the nutritive requirement of different livestock has been presented in the Ministry of Agriculture's Bulletins. These are created by the Technical Committee of the ARC of Britain. In these three separate reports, the requirements are set forth dealing with ruminants, poultry, and pigs.

Each of these reports summarizes the literature upon which the requirements are extensively based. Well, the most attractive feature of the British Feeding Standards is that the unit of energy requirements has been expressed in terms of Starch equivalent instead of T.D.N. or ME of NE are in Morrison and in N.R.C. standards.

5.3. USE/APPLICATION OF FEEDING STANDARDS

These days, feeding standards are discussed in various scientific meetings and tropical or temperate classrooms. Their significance for farmers is often doubtful, but it is then overlooked that even farmers apply feeding standards while feeding their farm animals (Figure 5.2).

Figure 5.2. Ensuring adequate diet to animals.

Source: *Image by Wikimedia Commons.*

However, the application of feeding standards by farmers is of different intensity, not mainly in the form of printed tables but usually with strict measures and objectives that vary from that of researchers.

Nevertheless, even under low input scavenging or pastoral conditions, the farmers apply their informal feeding standards to guide the farm animals towards better grazing grounds or away from unfavourable areas.

A lot of scientists, as well as development workers, are hesitant to apply the science of standards to the conditions of farmers. The reason behind this may be unfamiliarity with the concepts, or some other arguments as mentioned below:

- The fear of static and incorrect use of book values in certain conditions for which they were not developed. Shah et al. In the year 1980 made a common but incorrect statement on the application of feeding standards that "scientific feeding rates maximize milk production; they do not maximize the farmers returns from milk production." This statement basically means that feeding standards are more like a static rule that needs to be applied for a high level of production instead of just a flexible tool to aid common sense.
- Usually, the application of feeding standards is considered to refer to animals and feed alone, with inadequate attention to other components of the system that is crops eroding grazing areas, etc. As an alternative to applying the feeding standards to more generic problems, they are altogether abandoned.
- The use of feeding standards from one system to another can eventually lead to the inadequate classification of feeds as good or bad, without understanding that what is actually good in one system may be bad in another one.
- For instance, straw is considered as a bad feed for dairy animals in various high input systems; on the other hand, it is a practical feed to sustain animals through a dry season. In fact, straw is a vulnerable source of fibre for urban dairies where large quantities of concentrate are fed. In some of the urban feed or fodder markets of the tropics, the price per unit of DM for straw exceeds that of concentrate!
- The roughness of the performance prediction from standards in different situations make various workers forget the usefulness of the standards to predict the general trends in the feeding

system (McGraham, 1983; Preston, 1985). This eventually leads to endless testing of rations on the farm or station conditions. However, such testing cannot represent real situations, nor can it forecast precise responses owing to a lack of animals, etc.

Confusion regarding the requirement for feeding standards can be overcome by the definition and objectives of their use. The terms feeding standards and nutrient requirements can be interchangeably used.

According to Crampton and Harris (1968): "feeding standards are data sets which record what is believed to be the need of a specific animal for one or more of the recognized feed components for a given period of time."

The term 'for a given period of time' is used in place of 'daily' in order to include the time effects of compensatory gain or negative carryover from previous poor nutrition. Such data sets need not essentially be recorded in context to energy and protein values.

Standards that are present in the mind of farmers include criteria of food assessment, that actually relate to the effect of feed on butterfat content or drug consistency, for example (Rangnekar, 1991). Although, these criteria represent different aspects of nutritive value.

Additional refinements or modifications to the definition of feeding standards are possible, for example, the requirement of a particular animal can be further replaced by the requirement of a group of animals, etc.

Feed analysis can also be carried out by the farmers as it is done in the laboratory. Either way, it is done; it will eventually serve to establish the standards that would, in turn, serve to establish feeding or rationing systems.

This can be defined as the method of feeding in context to the allocation of quantity as well as the quality of the foodstuff over time, effectively guided by farmers, the availability of feed and the animals.

The utility and the required accuracy of feeding standards depend completely on the purpose and context for which they are being used. Scientists, farmers, and policymakers who develop their standards in isolation are likely to mislead each other.

Also, it must be clear that the use of farmers and scientifically developed standards of feeding are complementary instead of being mutually exclusive. Policymakers and scientists often fail to clearly understand the values and priorities of the farmers, especially in low input systems.

Many of the development policies operate on the assumption that 'high individual productions per animal would result in high income.'

However, this paradigm is not something new, as it is represented by the following quotation given by Jackson (198lb) – "the Royal Commission on Agriculture of 1927 first set forth the objectives of livestock development in India, namely increased production per animal and a reduction in animal numbers."

This paradigm requires to be combined in order to suit the objective of high production per animal, that is, a kind of approach that is only suitable for high input systems. Nevertheless, this does not mean that the feeding standards from high input systems are of no use in low input systems.

The discussion on feeding standards should concern the validity of data sets along with their application, not on the principle of their existence and application. To abandon the notion of feeding standards for its limitations would be to 'throw the baby out with the bathwater.'

Reassessment of the traditional classification of the tropical as well as temperate farming systems is needed for a better discussion of the feeding standards application. The technical validity of the feeding standards will be received along with other issues of their application in various socio-economic conditions.

5.3.1. Socio-Economic Aspects of the Use of Feeding Standards

If feeding standards were technically perfect, their use and application requires to be based on the local as well as seasonal feed availability, and also on the value of produce. Farmers in both low and high input systems can be presumed to produce for profit, however, economics is not the same everywhere.

Differences in demand patterns, resource allocation, social relations, job satisfaction, personal preferences, even within families and between different gender groups, eventually result in a wide array of production goals and systems (Harris, 1965; Crolty, 1980; Behnke, 1985; Macc and Houston, 1984; George, 1991).

Labour costs and produce prices vary even between different categories of farmers within the same geographical district (Patel et al., 1977). Low fixed costs or risk factors in combination with low produce prices force farmers to accept the low levels of production per animal (Doyle, 1974).

In various farming systems, the level of farm production is limited by production quota, whether openly in the high input system of the EC or indirectly in the low input systems under poor tropics conditions of marketing. In the case of production quota, income cannot be increased by a higher volume of production, just by a decrease in the cost per unit produced (Oscarsson, 1975; Welsch, 1975; Kristensen and Thysen, 1991).

Whether such a reduction or decrease is attained by the increased production per animal or by the decrease in the price of feed depends largely on the fixed costs of system-specific animal maintenance.

Objectives of multipurpose production (milk, meat, dung, young stock, wealth, status, and security) and hidden costs (disease risks, environmental side-effects such as erosion of common grazing lands, stress of entrepreneurship) make it a bit deceptive to calculate the profitability only in terms of a single output and input.

Eventually, such conditions allow no simple use of the earlier statement by the Royal Commission, even less so when animal production is secondary or integrated with cropping in mixed farming systems (Kidane, 1984).

Still, these micro-economic issues will ignore the requirement of reorienting the economic thinking for sustainability, for instance, to avoid externalization of hidden production costs (Conway and Barbier, 1990).

Economics are hard to calculate in systems where the production of milk is a by-product of the production of bullocks for the draft, where children tend the animals, or where feed is obtained from communal grazing.

Feeding standards have been largely criticized for not being applicable on farms. For instance, grazing animals will eat what they can harvest instead of eating a prescribed ration. Although, if grazing animals fail to grow and develop due to a suspected copper deficiency, then the knowledge of the copper requirement may be needed in order to confirm that the copper element is deficient.

In developing nations, most often, these standards are hard to be applied, the reason being the food required to supplement local resources are not easily available. While, on the other hand, in developed nations, there may be an inclination to oversupply nutrients that are cheap and readily available, like calcium.

Though, the need of avoiding oversupply by applying effective standards has been recognized in various countries where the excretion of extra nutrients by livestock causes environmental pollution like water and soil pollution in the Netherlands, for example, by nitrogen and phosphorus compounds. In future, feeding standards will probably prescribe the maximum as well as a minimum level of nutrient supply.

5.4. NUTRIENT REQUIREMENTS FOR MAINTENANCE

An animal is said to be in a state of maintenance when its body composition is constant, when it does not produce products like milk, and when it does not perform any work on its environment. As animals are kept rarely in this non-productive state, it might seem of academic interest to regulate the nutrient requirements for maintenance, although the total need or requirement of different classes of animal; mainly ruminants are factorially calculated by summation of the requirements for maintenance and production (Figure 5.3).

Figure 5.3. Welfare of animals by feeding them in open.

Source: *Image by Organic Sweden.*

As a result, knowledge of the maintenance requirements of animals is of theoretical as well as practical importance. The comparative importance of the energy requirements for maintenance is illustrated in Table 5.1. This shows the comparative contribution of maintenance as well as production to the total energy requirements of different classes of animal.

Table 5.1. Relative Contribution of Maintenance and Production to the Total Energy Requirement of Different Classes of Animals

	Net Energy Requirement (MJ)		**Maintenance (% Total Requirement)**
	Maintenance	**Production**	
Daily Values			
600 kg dairy cow producing 30 kg of milk	42	93	31
300 kg steer gaining 1 kg per day	23	16	59

50 kg pig gaining 0.75 kg per day	7	10	41
1 kg broiler chicken gaining 35 g per day	0.50	0.32	61
Annual Values			
600kg dairy cow producing 40 kg calf and 6000 kg milk	15,727	18,600	46
200 kg sow producing 16 (1.5 kg) piglets and 750 kg milk	7,100	4,600	61
2.0 kg hen producing 250 eggs	190	95	67

Source: *Table by Animal Nutrition, P. McDonald, R. A. Edwards, J. F. D. Greenhalgh, C. A. Morgan, L. A. Sinclair and R. G. Wilkinson, 2010.*

The animals illustrated in Table 5.1 are all highly productive. However, less productive animals proportionately use more of their energy intake for maintenance purpose. It may be estimated that, on average, cattle in Africa, for example, use around 85% of their energy intake for maintenance. Animals deprived of food are forced to draw on their body reserves to meet their nutrient requirements for maintenance. It is clear that fasted animals should catabolize the body reserves in order to provide the energy needed for key processes taking place in the body, such as respiration, circulation of blood, etc. As the energy utilization eventually leaves the body heated, animals then are in a state of negative energy balance. The same is true for other nutrients as well. For example, an animal consuming a protein-free diet will continue to lose nitrogen in its urine and faeces and is as a result come in a negative nitrogen balance.

The main aim of a maintenance diet is to prevent the drain on the body reserves. The maintenance requirement for a nutrient can be described as the quantity or amount needed to guarantee that the animal neither experiences a gain nor a loss of that specific nutrient. Therefore, the maintenance requirement is the minimum quantity that promotes zero balance.

5.4.1. Energy Requirements for Maintenance

Basal and fasting metabolism: as discussed above, energy used in the maintenance of an animal eventually leaves the body in the form of heat.

The quantity of heat evolved in this way is known as the basal metabolism of an animal, and its measurement provides a direct evaluation of the NE that the animal needs from its food so as to meet the maintenance demands.

The measurement of basal metabolism is further complicated by the fact that heat produced by the animal body not just from this source but also from the digestion as well as the metabolism of food components and the voluntary muscular activity of the animal. If the animal is kept in a cold environment, then heat production may also be increased further.

When basal metabolism is measured, the confusing effect of the heat increase of feeding is eliminated by depriving an animal of food.

When basal metabolism is measured, the complicating effect of the heat increment of feeding is eliminated by depriving the animal of food. The period of fast needed for the digestion as well as the metabolism of prior meals to be completed differs considerably between species.

In humans, an overnight fast is regarded enough, while on the other hand, in ruminants, absorption, digestion, and metabolism continue for many days even after feeding stops, and a fast of at least four days is required. The same period of fast is also recommended for pigs, while two days for fowl.

There are different criteria for establishing whether the animal has attained the post-absorptive state. If the production of heat is continuously measured, then the most acceptable indication is the reduction in heat production to a steady and constant level.

A second indication is given by the respiratory quotient. During the period of fasting, the oxidation mixture progressively changes from absorbed fat, protein, and carbohydrate to body fat and protein.

This replacement in the mixture of carbohydrate by fat is accompanied by the reduction in the non-protein respiratory quotient. When the theoretical value for fat that is 0.7 is reached, it can be presumed that energy is being acquired only from the body reserves.

In ruminants, a supplementary indication that the post-absorptive state has been reached is a reduction in the production of methane (CH_4), and thus the digestive activity is also reduced to a very low level.

The contribution of voluntary muscular activity to the production of heat can be declined to a low level when basal metabolism is calculated in human subjects, however, in farm animals, the cooperation required to get a state of complete relaxation can be hardly attained.

Fasting may restrict activity; however, even the small activity represented by standing as opposed to lying is enough for increasing the production of heat. As a result, in studies of farm animals, the term 'fasting metabolism' is used in reference to 'basal metabolism,' as strict basal conditions are not likely to be obtained.

Often a term is used in combination with fasting metabolism is 'fasting catabolism'; this incorporates comparatively small quantities of energy lost by fasting animals in their urine.

5.4.2. Influence of Climate on Energy Metabolism and Requirements for Maintenance

The effect of climate on the nutrition of farm animals is not just restricted to energy needs for maintenance but extends to other features of energy metabolism and also to nutrients other than energy. However, the climate has a major impact on the requirements of energy, and in cold climates, animals kept at or below the level of maintenance are highly affected.

Both birds, as well as mammals, are homeotherms. It actually means that they try to keep their body temperature constant. Animals continuously produce heat, and if they are to sustain a constant body temperature, they must lose heat to their surroundings.

The two key routes of heat loss are the so-called sensible losses by conduction, radiation, and convection from their body surface, and evaporative losses of water from the lungs and surface of the body (2.52 MJ/kg water).

In the first instance, the rate at which heat is lost depends on the temperature difference between the animal and its surroundings. In the case of farm animals, the rectal temperature which is a bit lower than the deep temperature of body lies in the range of 36°C to 43°C.

The rate of heat loss is also impacted by the characteristics of the animal, like insulation provided by the tissues and coat, and also by the characteristics of the environment, like relative humidity, air velocity, and solar radiation. In general, the rate of heat loss is determined by an intricate interaction of factors contributed by the animal as well as its environment.

5.4.3. Ruminants-Feeding Standards for Maintenance (Energy)

Ruminants: for maintenance, the energy requirements can be calculated from different feeding standards. However, it is almost impossible to include all of the standards. Requirements calculated by using some of the widely known standards are given below-

As published by the Agricultural and Food Research Council (AFRC) in the year 1993, the energy requirements for maintenance of cattle are based on fasting metabolism (F, MJ/day) and this can be predicted as follows:

Fasting weight is forecasted from live weight (W) by dividing it by 1.08, and metabolic weight is calculated by using the power 0.67 instead of more usual 0.75.

The fasting metabolism of bulls is regarded to be 15% higher than that of steers and heifers of similar weight, and an activity allowance (A, MJ/day) of 0.0071 W and 0.0095 W is included for the growth of cattle and dairy cattle, respectively.

Therefore, the NE requirement for maintenance (NE_m) of a 600 kg dairy cow would be calculated as follows:

If the ME content of the cow's diet was 11.0 MJ/kg DM, the efficiency of ME utilization for maintenance (km) would be 0.714.

The ME requirement for maintenance (ME_m) can be calculated as follows: $ME_m = 42.3/0.714 = 59.2$ MJ/day

As stated earlier, if an animal that is on a high plane of nutrition is fasts suddenly, then its fasting metabolism will be higher than that of a similar animal who was previously kept on a lower plane of nutrition.

As the estimates of fasting metabolism adopted by AFRC in 1993 were derived from beef and dairy steers fed at maintenance before fasting, they might undervalue the fasting metabolism of dairy cows fed at higher levels of production.

Evidence recommended that the fasting metabolism of modern dairy cows of high genetic merit, and on a high plane of nutrition may be considerably higher than that of previously recommended, and a value of 0.453 MJ/kg $^{W\ 0.75}$ has been adopted by fim (2004), which, based on their modelling approach, = 59.2 MJ/day

$ME_m = 42.3>0.714 = 42.3$ MJ/day

$NE_m = 0.53(600>1.08)0.67 + (0.0095 \times 600)\ F = 0.53(W>1.08)0.67$

Nutrient requirements for maintenance 355 give a fixed ME_m of 0.647 MJ/kg $^{W0.75}$.

The higher maintenance requirement of animals at a high level of production usually reflects the higher mass of metabolically active organs like the liver and gut. As with cattle, the maintenance requirements of sheep as published by AFRC in the year 1993 are based on fasting metabolism along with an activity allowance that differs between 0.0067 W for housed fattening lambs and 0.024 W for ewes on hill grazing.

The NE_m requirement of a 50 kg hill ewe can be projected as follows:

$NE_m = 0.23\ (50/1.08)0.75 + 0.024\ W = 5.3$ MJ/day

In contrast to AFRC (1993), where the requirements for maintenance are derived from calorimetric methods, the maintenance requirements of beef cattle published by NRC in the year 2000 are derived using the comparative slaughter technique.

This has the key benefit of allowing experiments to be done under conditions similar to those encountered in practice, and the impact of activity are subtly incorporated.

The NE_m requirement of beef cattle can be projected from empty body weight (EBW) as follows: $\mathbf{NE_m}$ = 0.322 EBW **0.75** (Where EBW = W X 0.85)

Here, the maintenance requirements are adjusted for the effects of breed and sex, with the requirements of Bos indicus breeds being declined by 10% and the requirement of dairy breeds being increased by 20%. The adjustment for effects of sex is almost similar to that made by AFRC (1993). In addition, certain refinements are also incorporated for the effects of climatic factors and nutritional status. It is quite interesting to note that the NE_m requirements of beef cattle were derived using the techniques of comparative slaughter and adjusted by 20% for dairy breeds, similar to those of dairy cows that were derived by using calorimetric methods.

5.4.4. Feeding Standards for Growth (Energy)

Ruminants: The UK Agricultural Research Council (ARC, 1980) assessed data on the body composition of a large number of cattle and sheep of different sexes slaughtered at different ages and weight.

AFRC in the year 1993 predicted that for cattle, the energy content of the gain made by castrated males of a medium-sized breed can be predicted as follows:

$Ev_g = (4.1 + 0.0332W - 0.000009W\ 2) > (1 - 0.1475\ LWG)$

Where; ev_g is the energy value of liveweight gain (MJ/kg); W is the liveweight (kg); and LWG is the liveweight gain (kg/day).

The first bracketed term in the equation defined the increasing energy content of gain as cattle grows in size, while on the other hand, the second term defines the correction for the increasing energy content of the gain related to higher live weight gains.

As a result, the energy value of the gain (evg) in a 100 kg animal gaining at 0.5 kg/day is projected to be 7.9 MJ/kg, on the other hand, the evg in a 500 kg animal gaining at the same rate would be 19.9 MJ/kg. The relative values for animals gaining at 1.0 kg/day are projected to be 8.6 MJ/kg and 21.6 MJ/kg, respectively.

To deal with the effects of sex and breed on evg, a simple 15% correction factor has been adopted. Therefore, for small breeds (early-maturing) and females the value projected is increased by 15%, while on the other hand, for large breeds (late-maturing) and males (bulls), the value is reduced by 15%.

A 500 kg female of a small breed growing at 0.5 kg/day would be projected to acquire $19.9 \times 1.15 \times 1.15 = 26.3$ MJ/kg. For sheep, the data analyzed by ARC in the year 1980 showed a large effect of sex on evg. However, only a small effect of breed (Merinos having more fat and hence a higher evg than other breeds), and no major effect of rate of gain. Based on these data, AFRC (1993) predicts that for sheep evg can be projected as follows:

- **Males:** evg = 2.5 + 0.35W;
- **Castrates:** evg = 4.4 + 0.32W;
- **Females:** evg = 2.1 + 0.45W.

The projected evg of male, castrate, and female lambs of 30 kg liveweight would be 13.0 MJ/kg, 14.0 MJ/kg and 15.6 MJ/kg, respectively.

5.5. MINERAL AND VITAMIN REQUIREMENTS FOR MAINTENANCE AND GROWTH

Minerals-Animals which are deprived of a dietary supply of mineral elements will sustain to excrete such nutrients. When elements are present in the body as constituents of organic compounds such as iodine in thyroxine, iron in hemoglobin, they are released from these compounds when they are worn out or expended (Figure 5.4).

Figure 5.4. Providing diet that have proper minerals and vitamins.

Source: *Image by feed Planet.*

The elements liberated are usually re-utilized within the body to a larger extent. Although, re-utilization is never complete and a proportion of every element will be lost from the body in the form of urine and faeces and also through the skin.

When elements are present in an organic form, like sodium, calcium, magnesium, and potassium, then the losses will take place in urine, like that of those arising from the maintenance of acid-base balance, and also in the feces owing to secretions into the gut that are not absorbed again. Due to such endogenous losses, animals need minerals for maintenance.

Losses of the endogenous mineral are usually small in context to the mineral content of the body. For an instance, a 30 kg pig that incorporates around 230 g of calcium losses approximately 0.9 g of calcium per day and as a result requires to replace around 0.4% of its body calcium.

In a similar manner, the same pig approximately contains 40 g of sodium and requires replacing 0.036 g or 0.09% per day. Comparatively, a pig would require replacing around 0.7% of its body nitrogen per day.

The techniques that are used in determining the mineral requirements are the same as those that are used to determine protein and energy requirements. A theoretical framework is provided by making use of the factorial approach, while the practical estimates of the requirements can be attained by using growth trials or nutrient balance.

As the mineral contents of foods are expressed as the gross or total amounts present, needs are stated in similar terms. Therefore, standards of nutrient should consider the differences in availability of mineral that occur between various species and age classes of animal.

Vitamins-there are no accurate estimates of endogenous losses on which vitamin requirements can be based. As a result, standards should be based on the feeding trails. As with the evaluation of mineral requirements, the key issue linked with the evaluation of vitamin requirements is establishing suitable criteria for adequacy.

Until today, the key criteria have been growth rate as well as freedom from signs of deficiency, as evaluated by physiological tests or visual examination of the animal, such as level of vitamins present in the blood.

In addition, vitamin storage can also be evaluated, either by analysis of tissue or from indirect evidence of tissue saturation provided by the excretion of vitamin in urine. The issues involved in evaluating requirements are given in Table 5.2. Table 5.2 shows that apparent requirements differ significantly, based on the preferred criteria for adequacy.

Table 5.2. The Vitamin A Requirement of Calves

Minimum Requirement for	Vitamin A (iu/kg live-weight/day)
Prevention of night blindness	32
Optimal growth	64
Limited storage of vitamin A	250
Maximal blood levels of vitamin A	500

Source: *Lewis and Wilson (1945). Journal of Nutrition.*

Practically, vitamin allowances should be high enough to avert the signs of deficiency and not limit the rate of growth. Although, higher allowances that promote storage or higher circulatory levels of the vitamin might be justified if they can be shown to improve the health as well as productivity of the animal, or the quality of animal products.

For an instance, high levels of vitamin E have been shown to increase the shelf life of meat. In most animals, storage is justified, as vitamin allowances are set at such levels that are enough for maintaining body stores, while both vitamin supply and requirements may vary.

The requirements of vitamin A and D of older animals, in general, are considered to be proportional to live weight, while those of the vitamin B group and vitamin E, which are more concerned with metabolism, differ with food intakes or with the intake of some specific nutrients. Therefore, the need thiamine which is specifically concerned with the metabolism of carbohydrate differs as per the relative proportions of fat and carbohydrates in the diet. Similarly, riboflavin requirements are increased by high protein intakes while; requirements of vitamin E are increased by high polyunsaturated fatty acid intakes.

However, the requirements may also differ, as per the extent to which B vitamins are produced in the alimentary tract. In the case of ruminants, microbial synthesis in the rumen may make the animal independent of a dietary supply.

Whereas in pigs and poultry, synthesis may occur in the lower gut, but the vitamins produced may fail to be absorbed efficiently. The contribution made by intestinal synthesis depends on whether the animals are free or not for practising coprophagy, which is defined as the eating of faeces.

Ultimately, requirements of the vitamin are usually determined by using diets with synthetic sources of vitamins whose availability may be higher than that of the natural sources in food. Nevertheless, very less is known about the availability of vitamin, a well-documented example of non-availability is provided by nicotinic acid in cereals, some of which is in a bound form that is not available to pigs.

5.6. FEEDING STANDARDS OF COW AND BUFFALOES

5.6.1. Feeding High Producers

There has been a physical limitation in terms of taking the volume which prevents the use of rumen. In computing rations, to get higher yields, it is impossible to achieve the same by just feeding bulky feeds or by intaking DM. This makes it impossible to achieve full energy.

Because of the limited space in the rumen, prior material cannot be reused unless it passes to the lower digestive tract. A mixture of fodder and grains is required to have enough food energy.

5.6.2. Frequency of Feeding

One cannot feed the whole grain in just one time, to result in it with high yielder as it can not only disturb fermentation in the rumen but also the environment is highly affected inside rumen. It would ultimately lead to the result of less production of the animal as well as deterioration in the health of the animal.

Certain parts of the same radon need to be feed on interval to overcome the effects of the high level of feeding. It's the case with both concentrates and roughage. Rumen fermentation are is reduced by feeding them twice a day, on the other hand, it would be eliminated if fed four times a day. This leads to proper utilization of proteins and improves digestibility.

5.6.3. Balanced Rations

Balanced rations are calculated to meet the requirement of actual feeding, which are calculated according to the feeding standards. All average animals have their own feeding schedule as compared to individuals, which might need more or less according to their feeding standards (Figure 5.5).

Figure 5.5. Ensuring balanced nutrition for cows.

Source: *Image by Food System Primer.*

Every animal needs to be taken care of about its diet whether the animal is losing weight or gaining the same and should observe the changes in the ration according to the same. One cannot always feed animals individually, as seen in a loose housing system where animals are being set free and are used to self-feeding habits, as one cannot feed roughage to them.

5.6.4. Guidelines to Feed High Yielders

- To obtain a good result feeding a ration that derives 30–40% of the feed units from grains and 60–70% from forages is required and it should include the optimum proportion of forage and concentrates in the ration.
- At the optimum stage forage is cut and it should be of excellent quality. The quality of fodder is affected only when there is a delay to cut the fodder.
- For high yielders, low-density high digestibility rations should be fed.
- Maintaining a continuous fermentation in the rumen is required, and for that, the schedule should be in such way.
- When high levels of grains are fed, feeding it mixed with the roughage or feeding it after the animal has consumed some forage.
- Feeding Cows in Late Lactation:
- Cows in late lactation are also in mid to late pregnancy.

A cow should be fed during this period for the following purposes:

- If she is a pregnant heifer, the maintenance of the cow and growth of the animal is required;
- Growth of the fetus;
- When she calves next, it is for production of colostrum; and
- Intake ability of the cow exceeds nutrient needs during late lactation, and to build sufficient reserves of nutrients in the body of the cow for the ensuing lactation.

5.6.5. Feeding Cows during Dry Period

One can define the dry period when an animal is in the advance stage of pregnancy and give more milk. Cows need to be feed in the rightly manner with high-quality green fodder which contains a high amount of vitamin A. Apart from this they also need to recover from deficiencies requiring salt, calcium, phosphorus.The growth of the fetus is accompanied by the formation of the membranes associated with it, and also by considerable enlargement of the uterus. In the early stage of pregnancy the amounts of nutrients deposited in the uterus and mammary gland are small, and it is only in the last third of pregnancy (from the sixth month onwards in cattle) that it becomes large. Mammary gland development takes place throughout pregnancy, but it is only in the later stages that it precedes rapidly.In a pregnant animal is given a constant daily allowance of food; its heat production will rise towards the end of gestation. The increase is due mainly to the additional energy required by the foetus for both maintenance and growth. It has been found that metabolisable energy taken in by the mother in addition to her own maintenance requirement is utilized by the foetus with comparatively low efficiency. The live weight gains made by pregnant animals are often considerably greater than can be accounted for by the products of conception alone. The mother herself, deposits 3 –4 times as much protein and 5 times as much calcium as is deposited in the products on conception. This pregnancy anabolism, as it is sometimes called, is obviously necessary in immature animals which are still growing, but it occurs also in older animals. Frequently much of the weight gained during pregnancy is lost in the ensuing lactation.

It has been seen that sources of calcium and phosphorous like bone-meal or di-calcium phosphate may be added to the ration when non-leguminous roughages predominate. If the soil and the crops are lacking in them then the other trace materials are required. There some more cases in which under the normal circumstances addition of such trace minerals is not necessary:

1. **Feeding Breeding Bulls:** A good quality of roughage and sufficient concentrates is required to fed the bull, it is required because to keep them thrifty and not fat. There is a requirement of an additional 2–3 kg of 13–15% protein concentrate mixture when the bull s receiving a liberal supply of good quality forage.

The maintenance of breeding bulls in good condition and suitable for breeding is highly essential requirement for the success of breeding programme. A rising condition is better for reproduction than a falling one. Fat males may produce semen of inferior quality or they may be slow or fail at service. It is sound practice to provide cool conditions and adequate drinking water.A balanced ration should be fed containing adequate energy, proteins, minerals, and vitamins. Green fodder must be available both before and during breeding season.It is of great importance that males should be fed regularly and not too much at one time, and too little at another.

2. **Feeding Young Stock:** The development of the rumen takes time and it is not developed by the time calf is born. To start functioning rumen will take a few months until rumen is fully developed. So, till then the calf is considered similar to a simple-stomached animal nutritionally.

It implies the following:

i. In the ration of calves, there is a requirement of essential amino acids, which should be provided in the required quantity. There are two sources of protein by the name high quality and mixed source by which they will get the benefit.

ii. Apart from vitamins A and D, vitamins belonging to the B-complex group are a nutritional necessity for them;

iii. Nonprotein nitrogenous substances like urea cannot be utilized by them;

iv. For the growth of bone, muscle, and other tissues, they require more minerals.

3. **Colostrum Feeding:** Through the colostrum the antibodies (gamma globulins) are transferred from mother to the calf. By the calf these gamma globulins will be absorbed and it will enter its system with the purpose of forming a ready-made antibody resistance system for the calf against all the disease, which will produce agents and other antigens the mother has had, met during its lifetime.

In the early stages, it is said that the calf against the diseases will be protected by them. And this will be done till their own antibody manufacturing system takes over. Apart from this colostrum is considered to be highly nutritious. It helps to prevent constipation and it is slightly laxative. The diet of the young calf should be of crude fibre, and the calf will also tend to constipate. Within half an hour after birth, calves must get colostrum.

5.6.6. Teaching a Calf to Drink From the Pail

It is said that for two to four hours, withhold feed to the calf and work should be done on an appetite. The process should be followed like finger should get a dip in milk and let the calf suckle from it.

Slowly lower your fingers to the pail containing milk pushing the calf's head if necessary, and by repeating the process, the calf will learn to drink from the pail. It should be kept in mind that the pail should be very neat and clean. According to the body temperature of calf milk should be heated. The calf can get scours by overfeeding.

Keeping the calf hungry is considered better than overfeed it. Similarly, milk with a high-fat percentage also causes scours. If the milk with a low-fat percentage is there then it can be fed but if the milk is with a high percentage of fat then it should be diluted with warm water and fed. According to its weight, the calf should be fed. For every 10 to 12 body weight per day, they require 1 kg milk.

5.6.7. Schedule of Feeding Young Calves

For 2 to 4 days the calf is first fed colostrum and then for a period of 10 to 14 days calf is fed with whole milk. Milk should be fed @ 10% of body weight in the first month

@ 15% of body weight in 2nd month if milk replacer is not fed

@ 20% of body weight in 3rd month if calf starter is not fed.

Then the next step is to mix the milk with skimmed milk, partially in the beginning and completely after two months along with an energy-rich concentrate mixture. Besides these, from the second week onwards, the calves may be fed on good legume hay or early cut green fodder.

Rumen development and establishment of proper rumen micro-organisms will be stimulated, and at an early stage, it will enable the calf to adapt to cheaper roughage feeding.

In India, a research is conducted in which the results say that feeding whole milk to calves beyond 2–3 months of age, as is done now, has no special advantage over feeling milk replacers and calf starters. There is an ideal feeding schedule for calves in general which can be prescribed and as follows for sustaining optimal growth:

- **Colostrum:** First 3 to 5 days;
- **Whole Milk:** 6th day to 40th day;
- **Milk Replacers:** 5th day to 60 days;
- **Calf Starters:** 2nd week to 90 days.

5.6.8. Feeding Bullocks

The requirement of an adult bullock has two components, viz.:

- The maintenance requirement; and
- The requirement for mechanical work.

It has been seen that the maintenance requirement is quite similar to the bullock and the cow and it is depends on the body weight. In muscular contractions, a portion of the nutrient energy is used, and the remaining is wasted as heat. Bodyweight has a major role to play, as a nutrient requirement for work production is totally dependent on body weight, type of work and duration. According to the purpose, the work may be light or heavy to which the bullock is utilized.

It is necessary to calculate nutrient requirements are for working bullocks for both normal and heavy work. Through roughages, nutrient requirement for maintenance can be met. Apart from this supplementation of energy-rich concentrates is a must for heavy work.

5.7. USEFULNESS AND LIMITATIONS OF FEEDING STANDARDS

5.7.1. Usefulness

Feeding standards comprise the quantity or amount of nutrients needed by an animal in order to remain healthy as well as highly productive. These standards are set in relation to productivity of meat, milk, eggs, wool, etc., the composition of the product that is the fat content of milk, and the physiological condition such as growth and foetal development.

Certain consideration is given to different regional conditions in the feeding standard. In addition, the feeding standards serve as a kind of guide for the feeding animals and also in calculating the efficiency of feed intake along with feed supplies for a various group of animals.

Usually, feeding standards allow profitable as well as effective nutritional management of animals suitable to the prevailing economic, nutritional, and sociological environment.

5.7.2. Limitation

Feeding standard cannot be considered as a complete guide for feeding as several other factors like the physical nature of the ration, palatability, etc. are not taken into consideration. These standards are approximately correct only.

The amounts of nutrients given in the table of feeding standards are proposed to be enough to meet the food requirements of maximum production. In addition, it will also be ineffective in changing environment as the environment may change the nutrient requirements. Feeding standards do not always consider the economics of livestock production; these are based on assigned nutritive values such as NE is confusing when the unconventional resources of feed are utilized. Feeding standards are not permanent; however, they are re-examined with the change in technology as well as the production targets.

5.8. CONCLUSION

At the end, it is concluded that the livestock sector is one of the most important sectors that provide food to a large number of people across the world. It is the duty of those people who deal with livestock to ensure that the feed gives to animals are up to standard. It is well known that if proper feeding standard is not maintained, then it can cause ecological imbalance. It is also seen that some animals diet contains a higher number of pesticides that may affect their health significantly.

In addition, as humans are also dependent on animals for their diet, they can also unintentionally takes these pesticides when they eat the animals. So, it is always important to keep a check on the diet that is given to the animals.

It is also seen that the feed needed by the animals depends on several factors such as their body size, sex, local environmental conditions, other animals with which they are living, their daily activities. So, all these factors should be in the mind of the dairy farmers while designing the feeding plan for animals.

Furthermore, now there are various technologies available in the market that can check the standard of the feed given to animals. Therefore, it is always better to make use of such technology in order to get benefit in the long term.

REFERENCES

1. AFRC (1993). Agricultural Food and Research Council. Energy and Protein Requirements of Ruminants. An Advisory Manual Prepared by the Agricultural Food and Research Council Technical Committee on Responses to Nutrients. CAB International, Wallingford, UK.
2. ARC (1980). Agricultural Research Council. The Nutrient Requirements of Ruminant Livestock. Technical Review by an Agricultural Research Council Working Party, Common wealth Agricultural Bureau, Farnham Royal, UK.
3. (No author name is given). *Feeding Standards*. [eBook] Jabalpur: Department of animal nutrition. Available at: http://www.ndvsu.org/images/StudyMaterials/Nutrition/FEEDING-STANDARD.pdf (accessed on 15 March 2021).
4. Behnke, R.H., (1985). Measuring the benefits of subsistence versus commercial livestock production in Africa. *Agricultual Systems*, 16: 109- 135.
5. Conway, G.R. and Barbier, E.B., (1990). *After the Green Revolution. Sustainable -agriculture for Development*. Earthscan Publications, London, 205 pp.
6. Crampton. E.W. and Harris. L.E., (1968). *Applied Animal Nutrition*. 2nd Edn. Freeman. San Francisco. CA, 663 pp.
7. Crolty, R., (1980). Cattle, economics and dcvelopmenr. Commonwealth Agricultural Bureaux, Slough. UK.
8. Doyle, C.J., (1974). Productivity. technical change and a peasant producer. a profiile of the African cultivator. Food Research Institute studies. Stanford University, Vol. XIII, No. 1. pp. 61-76.
9. EcoursesOnline.iasri.res.in. (2012). *Applied Animal Nutrition-I: Feeding Standards-Definition and Expression*. [Online] Available at: http://ecoursesOnline.iasri.res.in/mod/page/view.php?id=126314 (accessed on 15 March 2021).
10. Feed into Milk (2004). A new Applied Feeding System for Dairy Cows. Ed. by C. Tomas. – Nottingham University Press, 2004. – 68 p.
11. George, R.S., (1991). The female of the species: women and diarying ir, India. In: H. Singer and Prendergast (Editors), Development Perspectives in the 1990's. MacMillan. LJK. in press.
12. Harris, M., (1965). *The myth of the sacred cow*. In: A. Leeds and A.P. Vajda (Editors), Man, Culture and Animals, the Role ofAnimals in

Human Ecological Adjustments. Publ. 78. AAAS. Washington DC, pp. 217-218.

13. Jackson, M.G., (1981b). A new livestock development strategy for India. *World Animal Review*, 37: 2-8.
14. Kidane, H., (1984). The economics of farming systems in the different ecological zones of Embu District (Kenya), with special reference to dairy production. Ph.D. Thesis. Hannover University, Hannover, 227 pp.
15. Kristensen, A.R. and Thysen. I., (1991). Economic value of culling information in the presence and absence of a milk quota. *Acta Agriculturæ Scandinavica*, 41: 129-135.
16. Macc, K., and Houston. A., (1989). Pasioralist slrategies for survival in unprcdictablu environ- ments: a model of herd composilion that maxim& household viability. *Agricultural Systems*, 31: 185-204.
17. Makkar, H. P. S., & Beever, D., (2013). Optimization of feed use efficiency in ruminant production systems. *FAO Animal Production and Health Proceedings*, No. 16. Rome, FAO and Asian-Australasian Association of Animal Production Societies.
18. Makkar, H. P. S., Tran, G., Heuzé, V., & Ankers, P., (2014). State-of-the-art on use of insects as animal feed. *Animal Feed Science Technology*, 197: 1-33.
19. McDonald,P., Edwards,R. A., Greenhalgh,J. F. D., Morgan,C. A., Sinclair, L. A., & Wilkinson,R. G., (2010). *Animal Nutrition* (7th edn.) [eBook] Pearson. Available at: http://gohardanehco.com/wp-content/uploads/2014/02/Animal-Nutrition.pdf (accessed on 15 March 2021).
20. McGraham, N., (1983). Feeding standards, an outmoded concept in ruminant nutrition. In: D.J. Farrell and Pran Vohra (Editors). Recent Advances in animal Nutrition in Auslralia. University of New England Publishmg Unit. Armidale. pp. 184-191.
21. Morrison, F. B., (1917). The modified Wolff-Lehmann feeding standards. *Journal of Animal Science*, [Online] pp. 64–69. Available at: https://academic.oup.com/jas/article-abstract/1917/1/64/4744524?Redirectedfrom=PDF (accessed on 15 March 2021).
22. NRC, (2000). *Nutrient requirements of beef cattle*. 7th edition, National Academies Press, Washington D. C.
23. Oscarsson, G., (1975). The future of animal production: economic aspects. In: R.L. Reid (Edi- tor). *Proc. Third World Conference on*

Animal Production, 22-30 May 1973, Sydney, University Press. Thema 5, Pap. 57, pp. 447-457.

24. Patel, R.K., Kumar, P., Gangadharan, T.P., Dhaka, J.P., Vocgele, K., Sukumaran, N. R., and Sreekumaran, N . G., (1977). Economics of Crossbred Cattle. A study of the Cattle Breeding Programme of Indo-Swiss Project, Kerala. NDRI-Kamal, India.
25. Patil, P.V., & Patil, M.K., (2021). *Milk production management*. Jaya publishing house, Delhi, India.
26. Peeyusha, R., (n.d). *Feeding Standards of Cow and Buffaloes-Explained!| Livestock*. [Online] Your Article Library. Available at: https://www.yourarticlelibrary.com/livestock-management/feeding/feeding-standards-of-cow-and-buffaloes-explained-livestock/87109 (accessed on 15 March 2021).
27. Preston, T.R., (1985). Validity of feeding standards and development of feeding systems based on crop residues and agro-industrial by-products. In: T.C. Preston, Vappu L. Kossila. I. Goodwin and S.B. Reed (Editors). Better Utilization of Crop Residues and By-products in Animal Feeding: Research Guidelines. I. State of Knowledge: FAO Anim. Prod. Health Pap. 50, FAO. Rome, pp. 197-Z 13.
28. Rajesh, S., (2019). *Feeding Standards and Nutrient Requirements of Different Categories of Livestock and Computation of Rations-Part-1 | Pashudhan praharee*. [Online] Available at: http://www.pashudhanpraharee.com/feeding-standards-and-nutrient-requirements-of-different-categories-of-livestock-and-computation-of-rations-part-1/ (accessed on 15 March 2021).
29. Rangnekar, D.V., (1991). Farmers' perceptions of quality and value of feeds, fodder and feeding sys1ans. Paper presented at the lndo Dutch BIOCON Project Workshop Feeding of Rumi- nanls on Fibrous Crop Residues, Kamal, India, National Dairy Research Institute, Indian Council of Agricultural Research, Krishi Bhavan. Delhi, India.
30. Schiere, J., & De Wit, J., (1993). Feeding standards and feeding systems. *Animal Feed Science and Technology,* [Online] *43*(1, 2), 121–134. Available at: https://www.researchgate.net/publication/40185239_Feeding_standards_and_feeding_systems (accessed on 15 March 2021).
31. Shah. T., Tripathi, A.K. and Desai. M., (1980). Impact of increased dairy productivity on farmers' use of feedstuffs. Economic and Political Weekly, India, August 16. pp. 1407-1412.

32. Welsch, D.E., (1975). Resource use in systems of intensive animal production. In: R.L. Reid (Editor), *Proceedings of the Third World Conference on Animal Production*. Sydney, University Press, Thema 4, Pap. 52, pp. 409-415.

Chapter 6

Dairy Farm Management

CONTENTS

Dairy farm management is one of the most basic and important functions that determine the overall success of the dairy farm. In the chapter dairy farm management, the basics of managing a dairy farm are discussed. The chapter also explains the role & importance of dairy farm record keeping. In addition, the importance of record-keeping in dairy farm is also discussed.

It also shed some light on different types of record keeping in the dairy farm. It also demonstrated six critical principles in dairy herd management. The chapter also explains the importance of animal health management. At the end, some types of dairy cow diseases and disorders are explained in this chapter.

6.1. INTRODUCTION

It is generally seen that management is a key factor for the success of any business, and in dairy farming, the role of management is very imperative. Feeding, health management, and housing plays a very critical role in utilizing the full potential of dairy animals. The management practices comprise around 75 to 80% of the total cost incurred on milk production in the dairy business (Verma and Sastry, 1994).

Inadequate feeding practices of dairy cows results in poor growth, late conception, delayed maturity, and poor production. Provision of proper and comfortable spacing is beneficial in lessening the energy in preserving thermo-neutral zone and also provides comfortable, ideal, and hygienic conditions, which lessen the incidence of diseases, lower the pathogenic load, reduces the endo and ectoparasites and offers favourable environment for optimal milk production.

Dairy farm management should be complex, chiefly in the tropics with additional environmental stresses. Although, small scale dairy farmers often lack the essential management skills. Co-operatives are generally not strong enough to cope with proper health control and provide adequate services for their members. Herd management practices in cow handling, milking procedures, nutrition, housing, and sanitation play a chief role in prompting the individual animal as well as herds to diseases. Health management plays a very critical role as it helps in improving the conditions of dairy animals by lessening the disease load and right health status. Thus, having a comprehensive understanding of livestock management (housing, feeding, and health) practices would help dairy farmers in the identification of SWOT analysis (strength, weakness, opportunity, and threat) in livestock rearing.

This would further assist in the identification of adequate intervention policy to augment the production of dairy animals which will benefit the overall farming community. Although, records from smallholders are rarely analogous to those of western dairy nations. In the climatic, tropics, and environmental stress areas, especially heat stress, there could be a significant decline in animal productivity (Matthewman, 1993).

According to Akers (2002), a widely recognized lactation physiologist, the financing in milking management at farms where breed, feed, and animal care are wasted if milk handling and milking procedures are not pleasing. This suggests that the major area of focus should be on milking practice to promote good udder health and optimal milk production.

It is also important to note that dairy farming has been an important part of agriculture for several thousands of years, but historically, it was generally done on a small scale on mixed farms. The concept of large-scale specialized dairy farming was only applicable or feasible at a place where there is a higher demand for milk and its by-products and household people living nearby are dependent on the dairy farm for their milk consumption.

It was feasible as people usually buy dairy products such as milk, cheese, butter, cream on cash daily that allows dairy farmers to attain economies of scale.

Centralized dairy farming, as one can understand it by its name, is primarily developed around villages and cities, where residents were unable to have their own cows because of lack of grazing land and other factors such as lack of money and time to raise cows.

Near the town, farmers could earn some extra money on the side by taking care of extra cows and selling their milk to the people living in their town. The dairy farmers usually obtain milk in the morning and fill it in barrels and then bring it to market on a wagon. Before mechanization, the majority of the cows still milked by hand.

At the time of milking, they brought the vacuum pump and the automatic milking machine. The first milking machines were an extension of the conventional milk pail. The initial milker device installs on top of a regular milk pail and sat on the floor under the cow. After each cow being milked, the bucket would be dumped into a holding tank.

This resulted in the development of a Surge hanging milker. Before milking a CMV, a sizeable extensive leather strap known as a surcingle was put around the cow, across the lower back of a cow. The milker device as well as the collection tank hung beneath the cow attached with the strap. This innovation made it possible for a cow to move effortlessly during the process of milking rather than standstill over a bucket on the floor.

6.2. DAIRY FARM RECORD KEEPING

A dairy farm record is a file or document that is used to keep account of different events, activities, and materials with respect to the farm operations, while record keeping is writing or documenting those activities on recording machines or materials.

Record keeping is basically an essential element of good livestock business management. It plays an important role in taking financial planning decisions, providing valuable data and insights to the government administration, evaluating overall day-to-day activities in the dairy farm business, and assisting in taking decisions related to livestock management.

There are several types of dairy farm records a farmer should take into consideration in order to successfully run a farm business. These are records of identification of cattle, production records, financial records, a record of agricultural inputs and resources, health records, records of animal feeds, records of farm equipment and implements, dairy farm records, vehicle, and worker records.

Record-keeping can be attained through a selection of methods, from a basic simple manual (hand) record-keeping method to an elaborate computerized system. Selecting these methods depends upon a variety of factors such as the size of the herd, type of farms, the expected use of the records, and the persons who own the farm.

It is important to note that there is no "best" record-keeping system for all situations, but, at least, a farm records system should provide essential and precise information, fit into the organization or farm framework, and be always prevalent in order to assists in decision making.

Evaluating dairy farm record is known as record analysis. The evaluation process made it possible for the owner of a farm to make informed decisions based on actual farm performance. Therefore, establishing and using an efficient farm record-keeping system for existing farm activities and functions assists in farm planning, taking well-versed decision-making and analysis of both financial as well as production records.

Once there has been proper execution of a farm record-keeping system, analyzing the records can begin. Decision-making can be greatly augmented by examining both financial and production records and their subsequent impact on the revenue of the business. Better analysis can be done only if the data and information available on farms are complete and accurate.

There are a variety of factors a good record-keeping should take care of. Some of them are usefulness and leading to actions, having the potential to be able to convert into information, ensures simplicity, and avoid duplication as much as possible.

The first officially recognized recording system was witnessed in a dairy farm in the year 1985 in Denmark when milk recording organizations were formed. The recording movements disseminates promptly and now plays a critical role in all countries with an advanced dairy industry.

Several reasons that can impact the overall quality of farm records are the size of the farm, the education level of the owners of the farms, years of farming experience, and the status of the farmer. In addition, low numeracy levels and Illiteracy, lack of incentive, lack of time, lack of awareness are major restraints confronted by farmers in developing countries to keep dairy records.

6.3. IMPORTANCE OF RECORD KEEPING IN DAIRY FARM

Record keeping is a very important section in ensuring good livestock business management. With no written records, farmers need to rely upon their memory while making decisions with respect to their farm practices. But memories can become untrustworthy after some period of time; it may be a few days, months, or even years.

The ultimate objective to engage in a recording system is to enhance the level of herd performance by attaining things better, functioning of the farm more competently efficiently and effectively, or reducing the probability of poor future performances. Normally, Farm records are kept for some of the following reasons that are discussed in subsections.

6.3.1. To Assist in Financial Planning Decisions

Detailed and comprehensive financial records can be used for cash flow planning, farm analysis and other critical. Record plays a very important role to determine the performance of farming-related to the ventures which would be more profitable and which did not have an adequate return on investment and should be either made more efficient or eliminated.

It is often seen that farm managers use records to design balance sheets, producing cash flow and income statements, and assists in other financial prospects to make more informed decisions. According to a study conducted in a developing country India, it was found that identifying weak and strong points of business and providing the basis for farm planning are the key advantages produced by proper maintenance of farm record keeping. In a dairy farm, controlling labour is a significant element of financial decision making. This is generally a wages book recording wages paid, days worked, money owed, leave, and other related activities.

6.3.2. To Provide Data for Government Administrative and Extension Purposes

One of the most general and important uses of record in the majority of the developed countries is to ensure compliance with federal as well as state regulation and fulfil permit requirements. This is one of the imperative prerequisites of record keeping but should not be the primary reason, and a record system can be devised which satiates the receiver and is also essential for other objectives (Figure 6.1).

Figure 6.1. Collecting data of cow for various purposes.

Source: *Image by Medium.*

Some lending governmental bodies and related agencies require production and/or financial records to be maintained over a prolonged time. For instance, the government farm program necessitates certain acreage and production records be reported and retained by the owner of the farm owner. Record-keeping has implications for ensuring and fulfilling animal welfare and regulation of the welfare by the government. Recording gives a precise report on the adequate use of the antimicrobial product, including the rates, dose, and length of treatment, and withholding periods which is critical to address antimicrobial resistance. Progressively, farm owners are being required to maintain records about chemical use, irrigation water use, and livestock waste applications on their farms in order to be compliant with the environmental regulation.

6.3.3. To Assist in Livestock Management Decisions

Records of individual animals, as well as herds of animals, their health conditions, production, feed use, and other essential records, are imperative in assessing efficiencies by computing production per cow, Production per hectare, conception rate, feed cost per litre of milk, calving interval, stocking rate and culling rate are some of the important factors that help in taking decisions related to the management of the herd or individual. Several benefits that farmers gained by keeping records were: knowing expense, knowing income, managing revenue and expenditure more effectively, altering saving habit, engaging in communication more with family members, and trust more among family members.

6.4. TYPES OF RECORD-KEEPING IN DAIRY FARM

In the section, the primary area of focus is on farm production and animal health-related information (Figure 6.2). The keeping of farm production records can be categorized into five categories that are discussed as follow:

1. Animal identification;
2. Herd breeding records;
3. Individual cow milk production records;
4. Health records;
5. Feeding records.

Figure 6.2. Record keeping in dairy farm.

Source: *Image by APG Learning.*

6.4.1. Animal Identification

This information should be recorded only when there is an occurrence of any event. Putting this off lead would result in making certain errors and omissions. A proper and a well managed herd recording system will embrace the identification of each calf immediately after its birth.

In addition to the identification number or name, the birth date of the calf, its size, its sire and name, and the dam's names should be documented on a simple identification record. These kinds of information would be beneficial to dairy farmers in many ways when:

i. A sire is to be chosen and the farmer desires to avoid inbreeding;
ii. Heifers are assessed for breeding and superior bulls must be selected;
iii. Interpreting overall herd reproduction and identifying the age of heifers at first heat;
iv. Knowing the age at which a heifer should be targeted for breeding;
v. Evaluating whether the size of an animal is in accordance with its age;
vi. Comparing genetic lines across an area (with respect to other dairy farmers) to evaluate which animals to cull;
vii. Looked for the animal that should be culled based on their age.

6.4.2. Herd Breeding Records

A farmer with records based on recognized cows can advance his breeding management by being able to define such matters as the time at which to dry a cow off; finding the expected date at which cow can deliver a calf; highlighting bull services or poor insemination; establishing feeding programs and breeding dates; identifying sire, calf, and dams; and looking for the date for pregnancy testing.

Breeding records should be regularly updated and easily accessible. They should either be recorded in a diary or on an advanced farm on a computer and farmers must update it from time to time. There are several things that herd breeding records should include that are discussed as follows:

The identification of an animal including its birth date:

i.qName of sire and dam;
ii. Calving dates and comments;

iii. Heat dates and comments;
iv. Earliest breeding date;
v. Pregnancy examination;
vi. Information related to service;
vii. Expected date of calving;
viii. Drying off date;
ix. Any additional remarks.

6.4.3. Milk Production Records

Milk production records play a very important role to assists farmers in determining which animals are bringing down the average yield and therefore income from the herd. Better milk production records can provide a way to raise the production of milk from an individual cow and a herd through explicit management for individual animals. The more steadily and consistently individual daily milk production (and milk composition) data are recorded, the better it is for effective and efficient management. It would make the daily herd milk yields and milk quality/composition data easy to record and access. The seasonality of milk production needs to be calculated to develop the most adequate feeding strategies to address troughs in monthly milk yields.

6.4.4. Health Records

Calling to a veterinarian to assess animals once they witness health-related problem is a more usual tactic than arranging fixed veterinary visits. With the use of records, veterinarians would be able to get extra information about the possible causes of poor reproduction or ill health in an individual animal to offset for lack of familiarity with an individual arrival (Figure 6.3).

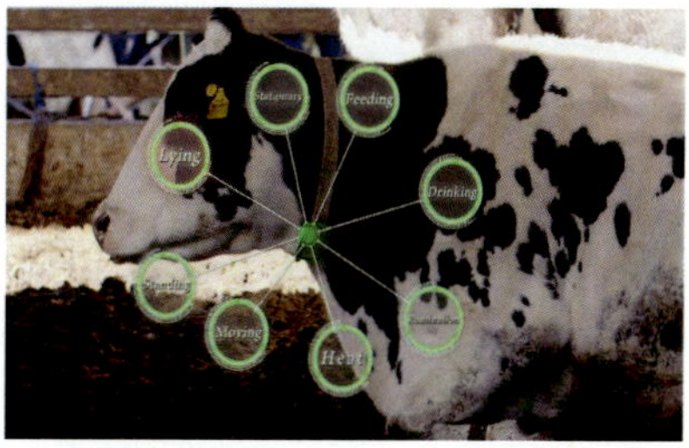

Figure 6.3. Checking health record of cow with latest technology.

Source: *Image by YouTube.*

Data or information concerning the production of milk, milk quality, dates of last change of status in lactation and reproductive terms, and other general information are of high importance along with the observation by farmers.

Health issues and treatment data can also be accumulated for individual stock on farm. As well as recording veterinarian visits (objective, date, diagnosis, and treatment), vaccination details (date and type) and cow and calf mortalities, records are valuable information in assisting disease management.

6.4.5. Feeding Records

It is important to note that feeding records can vary from simple records of when feed was bought and for which sort of stock, through to aspects of changes in feeding programs (such as daily quantities and types of feed per cow).

The more comprehensive the records of feedings are the better the capability to evaluate feeding efficiency (milk produced with respect to diet intake) on the farm. One of the efficient ways to record such information is to maintain a daily diary in the farm where changes in herd and feeding management can be updated daily. This diary could also be made into use to record o document appointments or other scheduled imperative farm activities.

An added aid to farm record keeping is a small pocket-sized notebook undertake by the farmer (and staff) in which they put data and information of any unusual observations, variations in routine management or other remarkable farm events.

Computers are now one of the most common ways to assists record keeping, interpretation, and collation, although a simple notebook and hand calculator would also do the work than relying on memory.

6.4.6. Action Lists

This term refers to details of specific stock demanding particular attention:

1. **Daily Action Lists:** These would embrace cows that are projected to show heat, cows due for drying off, through to cows and heifers due to calve.
2. **Weekly Action Lists:** These could comprise drenching of young stock, alterations in heifer and cow feeding groups, through to

heifers now prepared for insemination.

3. **Monthly Action Lists:** These may comprise the routine veterinary examination, such as heifers and cows for pregnancy testing. It would also embrace cows that have halted within the last month, reassessment of cows formerly treated by the veterinarian within the last month, cows not spotted to be on heat after 42 days from calving and also cows still not pregnant after 110 days post-calving.

Other stock that needs to be assessed would be cows with uneven cycles, cows confirmed in the calf but have cycled again and any other related issues noted in the daily diary. This would also include parasite and vaccination control necessities.

6.5. SIX CRITICAL PRINCIPLES IN DAIRY HERD MANAGEMENT

Dairy Herd Management is one of the essential factors for Dairy farms that are concerned with long-term high production. They are required to effectively manage many key factors: reproduction, cow nutrition, milking, and comfort. Alongside these prerequisites, all dairy farms should need to deal with the logistics of weather, expenses, and transportation.

It can be a lot to do with balance, but effective planning and insightful exploitation of the latest dairy farm technology and training make it possible. In this section, the principles of properly managing a dairy herd to ensure higher yields and satisfy cows are discussed.

6.5.1. Nutrition and Feeding (Figure 6.4)

Figure 6.4. Giving proper diet to cows.

Source: *Image by farmers Weekly.*

6.5.1.1. How Much to Feed?

One of the first and basic steps of planning nutrition is to understand how much to feed cows. It is important to note that cows are always required to intake higher nutrients in comparison with the milk they produced. When making decisions related to feed consumption, the industry standard is to measure food on the basis of dry matter (DM) content, or its original weight excluding water. Follow these guidelines for how much each category of cow needs:

- **Non-Pregnant Adult Cows:** 1.2% of the weight of body;
- **Non-Lactating, Pregnant Cows:** 2% of the weight of body;
- **Milking Cows:** 1.2% of the weight of body in addition to 5 kg per 10 liters of milk produced.

It is worth noticing that properly managing feed means cautiously monitoring the production of milk and bodyweight. If farm owners want your cows to reach peak lactation, they will need to make sure that they are feeding their cow's balanced high quality feed.

6.5.1.2. The Principles of Food Nutrition

The first principle of cow nutrition is to make sure that cows are getting adequate energy, which they'll get through fats and starch. Depending on the type of feed that exists in the region where the farm is located, that starch may come in the form of maize, wheat, sorghum, barley, or other grains. The fermentation of grain is what produces lactose within the milk.

Cows also need fat up to 6% of their ration. Sources of fat include brewer's grain, cottonseed, and oils. Overall, solid fats are good for the digestion of cows in comparison with liquids. Regular intake of vitamins is also required for cows.

Most fresh forage inherently comprises the minerals and vitamins that cows required at appropriate levels, but dairy farmers should also need to add supplemental vitamins during off-seasons when your feed is dried.

6.5.1.3. Selecting Forage

It is important to note that feed management comprises two main parts: storing silage and maintaining forage. For the majority of dairy farms, the grass is the cheapest as it is available in abundance and is considered as one of the finest quality forages available.

The primary objective is to keep pastures in a young vegetative state when it includes the most nutrients. This needs steady upkeep of irrigation, fertilization, and grazing rotation. Most farms will need to supplement natural grass during some seasons.

Farms having adequate grass quantity can dry their excess forage into hay. Otherwise, barley, maize, and legumes can all be good staple forages. The best choice will vary on the local availability-the most reasonable and good quality of crops varies from regions to regions.

Depending on the forage dairy farmers may choose, they would likely add concentrate and protein supplements, such as canola or soybean meal, citrus pulp, or rice bran. These supplements fulfil the gaps in the nutrients dried forage can offer.

6.5.1.4. Feeding Management

It is of utmost importance that cows should have steady access to clean water and superior quality feed. Each milking cows need at least 60–70 litres of water a day, plus an additional 4–5 litres per litre of milk produced.

To attain this, they need steady access to water troughs. It is often seen that cows drink a significant proportion of water immediately after their milking process; therefore, it is requisite to have an abundance of trough space as cows leave the milking parlour.

Similarly, fresh forage should also be steadily available. It must be noted that each cow should have more than enough space for forage access to make sure that dominant cows don't bully younger ones and thus stop them from eating.

In general, it is essential to feed with forage first, and then complement with nutritional concentrates. A basic rule is to feed 1 kg of concentrate for each 2 kg of milk a cow produces. This ensures that they are getting adequate nutrients.

6.5.2. Breeding and Reproduction

6.5.2.1. Goals and Planning

It is often seen that effective reproductive management requires effective planning. No matter how small or large, each dairy farm should indulge in daily monitoring activities for cows. Regular farm records will make it possible for dairy farmers to ascertain when cows will be in heat.

It means they known in advance which cows are pregnant, in heat, or are showing potential signs of any disease. Dairy farmers should also have proper plans in place for cows to undergo the evolution period and to tackle any health-related issues at the time of pregnancy.

6.5.2.2. Heat Detection-Using the Latest Technology

It is important to note that finding when cows are in heat is one of the most important steps in maximizing reproduction. Some farms just have basic knowledge of how to monitor cows visually. Standing to be mounted is a clear sign of being in heat. Cows in heat may also have the possibility of mucus discharge and become restless. This restlessness generates heat that can be identified by activity monitors. Much like human pedometers, activity can determine how much a cow walk. Peaks in activity can signify a cow in heat. Using pedometers on cows that are showing potential signs of heat can lead to a higher rate of heat detection, thus more efficient breeding. They can be particularly imperative on widespread farms, where it would be challenging for dairy farmers to visually monitor each cow.

6.5.2.3. Calf Health

It is often seen that one of the chief factors that affect calf health in the first initial days is whether they are kept in a clean environment. When calves are first born, they should be positioned in a place with a clean and dry environment. Afterwards, they can be kept either in small groups or individual pens. If dairy farmers want to put their calves in groups, then they should be separated into groups by age and size. One basic reason for this is to minimize access to manure, which harbours pathogens that can cause scours, the most common calf disease. It is also important to ensure that all the equipment that they are using for calf feeding should be kept in clean conditions, and all staff should wash their hands properly between handling adult cows and handling calves.

It is important to note that during the first 12 hours after birth, each calf should be fed 4 litres of colostrum, either directly through a bottle or through a tube to boost their immune system. In their initial few months, calves do not have fully grown rumens, and therefore they cannot digest hard food. They need to consume at least 15% of their body weight in milk every day. During their first week of birth, dairy farmers can introduce high-quality hay and concentrates so that calves initiate to grow their rumen and get used to forage. This will help in preparing them for weaning.

6.5.3. Milk

6.5.3.1. Milking Tips

It is worth noticing that milking is never as simple as joining cows to the equipment. Some of the tips that would be useful in making the milking process as efficient and low-stress as possible are discussed as follows:

- Cows are often vulnerable to stress-farm should always try to give them quiet and a calm environment;
- When moving the cows into the milking parlour, keep cows walking slowly to prevent slippage;
- Teats should be dry and clean before connecting to milking equipment;
- Antiseptic should always be put on teats after the milking process to prevent any potential disease;
- Milking equipment should always be monitored regularly for checking any blockage in vents and watching for the flow of milk;
- Whenever the milking process is done, its equipment should be cleaned with sanitiser and water.

6.5.3.2. Managing Milk Quality

It is important to note that immediately after the milking process; the milk needs to be chilled. Dairy farms those are larger need to have their own refrigeration units. Otherwise, the milk needs to be conveyed immediately after the milking process to the cooling units to prevent any contamination.

6.5.4. Cow Welfare

If the farmers want high production, then they need to provide a comfortable environment for cows. Ensuring a comfortable environment is required to keep in mind such things (Figure 6.5):

- Clean, dry bedding;
- Shade in the heat;
- Room to lie down;
- Non-slip flooring indoors.

Figure 6.5. Ensuring welfare of cow.

Source: *System Nextgen System.*

6.5.4.1. Shed Design Options

It is generally seen that the ideal dairy barn relies on the region, chiefly the weather. A dry manure lot may work efficiently in a dry region, as it offers enough space and is also economical to set up. Although, rainfall creates mud that considerably enhances the chances of illness. Compost barns feature sawdust or other bedding material spread out in a large shed.

As long as farmers have adequate and economical bedding material available, compost barns can be a comfortable and reasonable design. Free-stall barns are pricier to construct but deliver some benefits. Free-stall barns facilitate individual cubicles along the sides with a broader open space in the centre.

Cows can freely arrive at the stalls to eat or lie down. This setup not only gives cows more space to roam but also make it easy for dairy farmers to manage to feed and to keep manure separate from the bedding.

6.5.4.2. Optimal Shed Design

Wit is worth noticing that while designing a cowshed, there are many key areas that dairy farmers should take care of. Although it may be a bit pricey to enhance these details, they will pay off in terms of overall productivity and better cow health:

- Stalls with adequate space for cows to take the sides and move forward;
- Proper ventilation facility;
- Non-slip flooring outside of the stalls;

- Effluent management that keeps manure separate from bedding and food;
- Space for feed;
- Soft bedding that is fluffed regularly and substituted repeatedly;
- Access to good quality food and clean water.

6.5.4.3. Veterinarian Care

Veterinarians will be an essential element in managing cow health. While even if dairy farmers are spending enough amount of time in managing many parts of cow welfare, they need to take the veterinarian activity of cows seriously as it is concerned with providing medication, performing testing, addressing medical emergencies if any, and certifying essential slaughters.

It is essential to include veterinarian bills into the budget of dairy farming, and preferably farmers need to establish a relationship with a local vet they like. The time when a group of cows need treatment or antibiotics for mastitis, then the dairy farmers will have a team of professionals ready whom they can trust.

6.5.4.4. The Economics of Herd Health

As it is well known that veterinarian care can be expensive, so does the health issues. These range from the declined production of a lame cow, to the loss of usable milk from mastitis, to the death of sick calves and cows. It is often seen that majority of the dairy farmers spend most of their time on disease treatment rather than on preventive care.

Some disease is unavoidable, but high preventive care can often lower the possibility of disease and be most cost-effective in the long term. Preventative care comprises obvious treatments, such as vaccines. Many vets also offer training, advising, and consulting on cow health. There is a need to take this type of preventative care in a budget and consistency can advance profitability and cow health over time.

6.5.5. Heat Stress and Comfort

6.5.5.1. The Right Temperature

It is generally seen that for the majority of cows, an environment temperature falling between 6 and 18°C, or 43 and 65°F is perfect. Especially during the summer, high temperatures may impact the production of milk and consequently lead to heat stress. Cows are not very successful at circumventing the heat-they do so typically through breathing.

During hot months, it is thus particularly vital to manage humidity levels and to give cows plenty of space. It is important to note that the comfort range of cows will vary based on breed, and some cows have been bred exactly to withstand colder or hotter temperatures. Proper planning should be based upon the category of the breed.

6.5.5.2. Cooling Strategies

It is usually seen that there are several different tactics that can help keep the temperature cool, and these strategies initiated with shed design. First and foremost, cow sheds need proper ventilation. Open-sided sheds can be opened up in the summer in order to expand ventilation scope, and the top of the roof should also be opened to keep air flowing.

A steeper roof would also be useful in enhancing the flow of air. Outdoor pastures should always have adequate shade to make sure that cows will get sunlight at least once a day. When ventilation alone is not enough, fans and sprinklers can keep cows cool. Ceiling fans within a shed work to increase evaporation and ventilation.

Sprinklers are conceivably the most efficient modes to keep cool cows-they wet cow's backs, and the evaporation of the water cools them down. The majority of the farms find it productive to install sprinklers above the feed stations within the holding yard. That way they can remain wet and have adequate time to dry before milking.

6.5.6. Dairy Industry Innovators

Technology may be an investment that requires huge setup cost but is essential for the long term as it would help reduce running costs and enhances productivity over a while. Looking for the best technology is an utmost important task and should be taken vigilantly to choose innovating technology designed for effective dairy herd management.

Some of the technologies and mechanical tools that would be helpful in effective and efficient management in the farm are pedometers and milking robots, as well as the software used to help plan and manage everything from milk quality to herd health.

6.6. ANIMAL HEALTH MANAGEMENT

In this section, the primary objective is to explain good dairy farming that is important to order to ensure animals that produce milk are healthy and there is an effective health care program for cows.

Although, it is important to note that that of the practices are pertinent in all situations and may be surpassed by the national, market as well as international demands. The recommended suitable dairy farming practices for animal health are set out under the following headings:

- Establish the herd with resistance to disease;
- Prevent entry of disease onto the farm;
- Have an effective herd health management program in place;
- Use all chemicals and veterinary medicines as directed.

6.6.1. Establish the Herd with Resistance to Disease

6.6.1.1. Choose Breeds and Animals Well Suited to the Local Environment and Farming System

It is generally seen that different dairy species and breeds have distinctive requirements. Certain dairy animals that are well adapted to the local environmental conditions are more likely to have minimal risks of productivity posed by animal welfare and health problems.

Of specific significance is the animals' capability to adapt to climatic extremes, local parasites (particularly ticks), feed quality, and their learned resistance to endemic disease. The demands on the animals are also dependent on the farming system adopted by the dairy farmers.

Housed animals may likely have greater exposure to communicable diseases, whilst considerably maintained animals are more vulnerable to parasitic infections. Animals that are conveyed from different locations may be more susceptible to endemic diseases in the new location because of insufficient past experience and the development of immunity.

6.6.1.2. Determine Herd Size and Stocking Rate-Based on Management Skills, Native Conditions and the Availability of Infrastructure, Land, Feed, and Other Inputs

Larger herds and higher stocking rates usually need a higher level of infrastructure, organization, and skill to manage. The risks are augmented, especially in the standardized dairy farming systems. Disease burdens can be significant and individual animals needing intervention can be more challenging to treat and identify. There is always a need for good planning and management skills to circumvent the risks that have severe implications.

In developing countries, droughts, overstocking, and fires should be taken into consideration while determining stocking rates. Drastic seasonal changes (for instance, deep snow, Monsoons, high humidity, subzero temperatures, or heat) may require distinctive farming systems (e.g., pastures or houses) contingent on the time of year.

Fodder supplies should be planned cautiously for all possibilities as dairy animal's need a persistent source of superior feed quality and water at all times.

6.6.1.3. Vaccinate All Animals as Recommended or Required by Local Animal Health Authorities

Vaccination is generally considered a very important tool to restricts the impact of disease by raising the immunity of the animal population to definite pathogens. Animal health authorities should update local dairy farmers from time to time about the information related to specific vaccines which are suitable for a particular area. In some dairying regions, dairy farmers need to comply with the local law to vaccinate their stock against serious contagious diseases.

6.6.2. Prevent Entry of Disease Onto the Farm

Only buy animals of known health status (both individual animals as well as herd) and control their insertion to the farm by using a quarantine period if signified. One of the best and finest ways to prevent the spread of any infectious diseases is to keep a group of farm animals as close as possible.

It is important to make sure that no new animals take an entry into the group and erstwhile resident animals do not re-enter after they have left the herd. It seems easy, but in reality, is challenging in practicality, so there is strict control and vigilance if any new animal introduced onto the farm. There is also a high probability of risk of disease if a group of animals share grazing or other facilities.

Before being introduced to the farm, all dairy animals and herds should be inspected for any possible diseases that are noteworthy to their area of origin and new location. All animals should have:

- An identification system in order to make tracking possible in relation to their origin (birth to death identification system).
- Some form of vendor certification or declaration that details the disease/health status of animals and any suitable tests, vaccinations, treatments, or other practices that have been or are being carried out.
- It is important to note that potential sellers of dairy livestock should keep proper permanent animal health records for their animals.
- The health status of the vendor herd should also be properly documented. This is essentially imperative for diseases having extended incubation periods such as paratuberculosis, etc.
- When the status of the animals' health is unknown, they should be positioned under quarantine for some time or kept away from other animals for a certain period.
- Sick animals should be prohibited. It is always preferable to consider treating all hosted animals for internal parasites on the entrance. Keep and maintain up-to-date records of all animal movements to and from the farm.

6.6.2.1. Ensure Animal Transport on and Off the Farm Does Not Introduce Disease

It is always important for new or potential buyers to ask for a detailed report of live animals to determine if the animals are diseased or sick. It is recommendable to make sure that no infirm or sick animals should be conveyed alive. An aptly skilled veterinarian or operative should carry out any euthanizing needed on-farm. The disposal of dead or diseased animals should be carried out in such a way that lessens the menace of the disease being spread out and in line with the guidelines framed in the OIE Terrestrial local regulations and/or Animal Code. For instance, transport vehicles should not convey diseased or dead animals from one to another farm, without taking adequate measures to curtail the risk of spreading disease.

6.6.2.2. Monitor Risks From Adjoining Land and Neighbours and Have Secure Boundaries

It is always suggestible to be aware of any exotic diseases and/or local (endemic) diseases that possess the ability to impact the wellbeing of the flock or herd, particular from neighbouring farms. Place animals in such a manner to make sure that there is no risk of disease being spread between farms and within farms.

6.6.2.3. Where Possible, Limit Access of People and Wildlife to the Farm

It is often seen that people, as well as vehicles that roam between a numbers of farms, have the highest possibility of spreading the disease between the farms. Therefore, always keep milk/tanker pick up access and public tracks clear of faecal contamination.

Restrict access to an 'as needs' basis and put in place adequate processes to abate the spread of disease. On a farm, when visitors are entering, they should wear clean protective clothing as well as disinfected footwear if arriving areas that pose a greater threat of conveying disease onto or from the farm.

There should always be a proper record of all the visitors that have visited the farm even once in their lifetime. It is important to note that disease can be spread both from as well as to humans and wildlife. Avoid visitor contact with animals as much as possible.

6.6.2.4. Have a Vermin Control Program in Place?

Make sure that adequate vermin controls are in place across the entire place where vermin could breed introduce disease and/or affect the overall quality of milk and safety. Vermin breeding sites should be eradicated, particularly if those sites also harbour disease pathogens, such as livestock disposal sites, manure heaps, etc. Vermin control measures may also be needed at a place such as feed storage, milking shed, animal housing areas, and water storages. Vermin species differ from location to location but can comprise of indigenous animals, birds, rodents, and insects.

6.6.2.5. Only Use Clean Equipment From a Known Source

it is important to make sure that all the veterinary and agricultural equipment presented on to the farm is clean and adequate measures have been in place to avert the introduction of disease. It may comprise of giving answers to certain questions such as the past history from where the equipment comes and how it has been used. Take extra care with borrowed or shared equipment.

6.6.3. Have an Effective Herd Health Management Program in Place?

6.6.3.1. Use an Identification System That Made it Possible for All Animals to Be Identified Individually From Birth to Death

All dairy animals should be simply recognizable by all people who come in contact with them. This system should be permanent in place, allowing all the animals to be distinctively recognized from birth to death. Examples of identification systems include tattooing, ear tagging, radio frequency identification (short form: RFID) measures such as freeze branding, and freeze branding.

6.6.3.2. Develop an Effective Farm Herd Health Management Agenda Aimed at Prevention That Fulfil the Needs of Farm as Well as National and Regional Requirements

It is the ultimate objective of herd health programs to keep all livestock productive and healthy. It comprised of the farm's practices for the diagnosis, prevention, treatment, and control of pertinent animal diseases, including external and internal parasites.

It is imperative to ensure a reliable tactic to herd health, so all staff should be aware of and comprehend the farm's herd health program. The program should encompass all facets of animal handling and husbandry, milk harvesting as well as other practices of dairy farm management concerned with the health of animals.

This may include disease screening, control measures and/or vaccination being needed by supply contracts or animal health authorities. Where effective vaccination is present, they may be made into use to enhance disease resistance.

Prophylactic treatments may be required as protective measures when no other suitable options present. Effective herd health programs should be established in discussion with suitably skilled people like veterinarians.

6.6.3.3. Regularly Check Animals for Signs of Disease

It is always important to keep an eye on all animals frequently and use confirmed methods to assists in exposure and precise diagnosis of infectious disease. Some useful tools may comprise of rectal thermometers, observations of body condition, and animal behaviour, and examination of foremilk. Laboratory or other relevant tests may be essential to screen animals for disease.

Herd and/or animal-level disease testing may also be accessible through herd improvement/communal milk collection centres or statutory disease control programs.

Detailed reproductive and breeding records should be kept and animals spotted at adequate stages as many diseases are allied with reproduction. Clinical diseases should be examined to identifying the underlying cause(s) so that animals can be treated and further cases averted. Some consistent management practices such as hoof care programs would be useful in curtailing the incidence of lameness.

6.6.3.4. Sick Animals Should Be Attended Speedily and in an Adequate Manner

Treat all types of injury, disease, and poor health by confirmed methods after precise diagnosis. Treat diseased animals suitably to curtail the occurrence of infection and the cause of pathogens.

6.6.3.5. Keep Sick Animals Isolated

Where possible and if appeared, keep sick animals isolated for as long as possible onto the farm to curtail the spread of contagious disease. Provide separate facilities and/or milk sick animals last.

Prompt treatment can restrict the extent of infectious agents. Disinfect and clean equipment after they were used for sick animals and ensure people who are coming into contact with these animals should take adequate measures to avoid infections.

6.6.3.6. Separate Milk From Sick Animals and Animals Under Treatment

It is always suggestible to follow adequate procedures to isolate milk from sick animals and animals which are under treatment. This milk is not adequate for consumption by humans and if stored on-farm should be labelled clearly by the same. Clean milking utensils and equipment comprehensively to evade cross-contamination.

6.6.3.7. Keep Written Records of All Treatments and Identify Treated Animals Appropriately

Staff, veterinarians, and others concerned with managing dairy animals on the farm must know what treatments would be apt for which animals. Put in place an adequate system to willingly recognize treated animals, and record proper details with respect to the local regulations and to manage withholding periods for meat and equipment.

6.6.3.8. Manage Animal Diseases That Can Affect Public Health

It is always advisable to follow local regulations and OIE recommendations to control zoonoses. Aim to keep diseases of public health importance at a level in animal populations that is not dangerous to people. Always try to evade direct transmission to people through the right animal supervision and hygienic practices. Always try to safely dispose of carcasses and animal waste.

Prevent the adulteration of milk with urine and faeces or other animal wastes. Never use milk taken from sick animals for human consumption. Manage the risks professed by drinking raw milk from farms.

6.6.4. Use All Chemicals and Veterinary Medicines as Directed

6.6.4.1. Only Use Chemicals Permitted for Supply and Use Under Pertinent Legislation

Only use those chemicals that have been registered and assessed for use in dairy production by the concerned authority.

6.6.4.2. Use Chemicals According to Directions, Calculate Dosages Carefully and Observe the Appropriate Withholding Period

Always use veterinary and agricultural chemicals for the objective for which they were approved, and in agreement with label directions, gives an expectable outcome whilst handling the potential risks.

Dairy farmers should carefully use all the recommended chemicals to prevent the chemicals from negatively impacting animal health and productivity, the safety and health of the user, the environment, or the quality and safety of meat and milk products. Always be cautious about chemicals that can leave residues in milk. These may contain teat disinfectants, detergents, anti-parasitic, dairy sanitisers, antibiotics, pesticides, herbicides, and fungicides.

Dairy farmers should:

- Use chemicals only for the objective for which they are intended at. For instance, lactating animals should never be treated with veterinary products which are not apt for the treatment of animals producing milk contributed for processing or otherwise used for human consumption.
- Always read the label as it may contain some vital information related to the legal and safe use of the chemical.
- Always follow the advice written on the label and any risk assessment or chemical datasheet.
- Observe the specified withholding periods.

6.6.4.3. Only Use Veterinary Medicines as Prescribed by Veterinarians

It is often seen that veterinary medicines pose risks to animals, humans, and food safety and thus subjected to restricted control on their supply and use. Therefore, it is always recommended to use approved veterinary medicines, at the appropriate dosage as directed in the label, or as advised or prescribed by a veterinarian. Relevant withholding periods must be observed.

All veterinary chemicals and medicines aimed at treating food-producing animals should have a withholding period specified on the label. If label directions are not followed strictly, the stated withholding period will not be applicable.

If no labelling instructions printed or no withholding time is specified, it is recommended to not use the product. The use of veterinary medicines in contrast to the label recommendations is dubbed 'off label use' and poses added threats. Off-label use of veterinary medicines should only be used under the supervision of veterinarian and in compliance with regional/ national regulations.

6.6.4.4. Store Chemicals and Veterinary Medicines Securely and Dispose of Them Responsibly

It is recommendable to store veterinary medicines and chemicals securely to make sure that they are not used improperly or do not inadvertently contaminate feed and milk. Always look for the expiry date of a product before buying it. Chemicals and their containers should also be discarded in such a way that do not come in contact with the animal as it may contaminate it as well as the farm.

6.7. TYPES OF DAIRY COW DISEASES AND DISORDERS

It is important to note that dairy cows are susceptible to infection and several types of disorders. Therefore, every farmer should consistently monitor the animals to address and detect the problem if any, at the earliest stage (Figure 6.6).

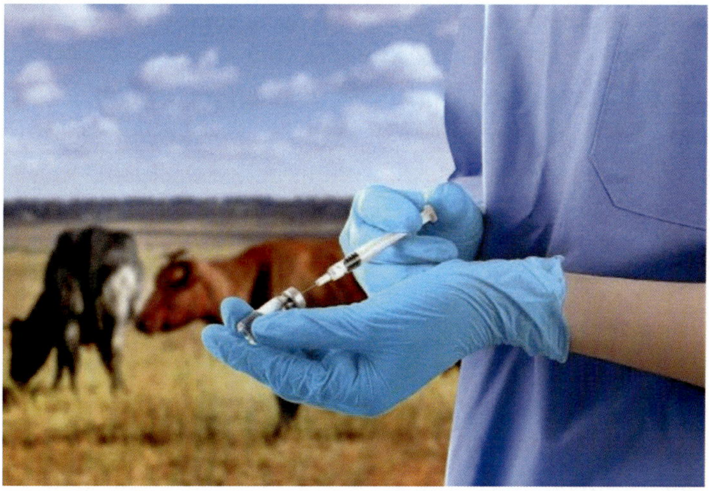

Figure 6.6. Giving proper treatment to cows is very crucial these days.

Source: *Image by Teplu.*

Although several problems can be evaded by vigilant management or dealt with on the farm. Every dairy farmer should have contact with a veterinary doctor that can be called immediately whenever a serious problem arises. Some of the common problems that can happen to dairy cows are discussed as follows:

1. **Acidosis: (Engorgement Toxemia, Acute Indigestion):** The rumen may develop an acid condition (pH of 4.0 to 4.5) that affects normal function and digestion of rumen. Avoid unintentional access or speedy changes to a high-energy feed, for

instance, grain mixture, or highly moisture corn. In severe cases, early diagnosis and treatment are very significant to save the life of animals.

2. **Bloat:** It usually happens due to excessive gathering of gases in the rumen. Severe bloat can arise after feeding on legume pastures. Breathing becomes difficult and excessive salivation is usual. The cows' left side generally balloons. Therefore, it is recommendable to feed cows on dry hay before permitting grazing or give cow's appropriate bloat preventing drugs.
3. **Foot Rot:** This arises by a break in the hoof or skin, generally between the toes, letting bacteria take an entry. Symptoms are a quick, swollen foot; progressive lameness; and a characteristic foul odour. Infection may convey to joints, or even disseminate to the leg, eventually killing the cow.

In a way to avert this problem, yards should be washed of any things that might cause a break in the hoof or skin. Soft, non-callused feet are extremely vulnerable to injury. Small stones rooming between the toes can also be a drawback. Animals should be delivered with the suggested zinc and iodine levels. Use a footbath with copper sulfate (2%).

4. **Grass Tetany (Hypomagnesemia):** It is likely the case with cows grazing on lush grass pasture which has a high proportion of nitrogen, resulting in low absorption of magnesium. It will cause cows to abruptly develop tetany, walk with a stiff gait, and go into convulsions, fall, and die.

Cattle that graze on grass fields that are fertilized with a high proportion of nitrogen should be monitored carefully and supplemented with 60 gm of magnesium oxide daily during this period to compensate for the harm. Grass tetany could be occurring in calves fed only milk and green lush pasture.

5. **Foreign body syndrome or Hardware Disease:** It causes by a puncturing of the reticulum if a cow swallows a sharp object. The animal will have a sudden lack of appetite, a careful gait, and an unwillingness to move. Respiration is recurrently fast, pulse rate is abnormal, and rectal temperature is 40°C or higher. It is, therefore, recommendable to give magnets to cows when a herd problem is prevalent. Careful auscultation of the thoracic region and x-ray examination provides a confirmatory diagnosis. If there is a penetrating foreign body in the reticulum, the rumenotomy operation is a must to remove the penetrating foreign body.

6. **Mastitis:** It is the infection caused in the mammary gland by any one of several bacterial organisms. Symptoms vary with the range of inflammation. Acute cases show a painful and swollen udder and often cause the cow to devoid hunger.

Chronic cases of mastitis have somewhat swollen udder quarters and small flakes in milk. At present, there is no feed/supplement available in the market that can cure mastitis. Although, an unexpected addition of nutrients to the feeds may result in increased production at the same time will also increase stress among mammals. This, in turn, might cause subclinical (below level of detection) cases of mastitis. To prevent mastitis, it is advisable to consult a veterinarian. Feeding suggested levels of vitamin E and selenium might be useful. Adopting loose housing system for animals, clean and dry animal shed and premises along with balanced feeding prevents mastitis.

7. **Milk Fever (Parturient Paresis):** This condition usually arises at calving and is initiated by an abrupt deficiency of blood calcium. Initial sign is staggering, followed by the struggle in rising, and lastly falling and being unable to rise independently. Cows are generally down with head turned back towards the flank. Late action may cause the death of a cow or it may develop a slow response to treatment.

This problem can be curtailed by feeding low phosphorus (30 to 40 gram) and calcium (less than 100 gram) ration during the dry period. Cows that are vulnerable to milk fever should be fed on an exact calcium short ration 10 to 14 days before calving, or add anionic salts to the ration.

Anionic Salts would make the diet more acidic, improving blood calcium levels by exciting calcium to convey from the bones and enhance calcium absorption from the small intestine to the blood and body.

8. **Moldy Feed Toxicity (Aflatoxins):** The Aspergillus *flavus*, fungus, as well as certain other moulds, may yield toxic substances when feed grains are stored under poor ventilation or high moisture conditions. Cows that are given such feed might develop large adrenal glands, fatty liver degeneration, and oversized bile ducts.

It would result in less intake of feed, lower production of milk, and or poor reproductive performance. In this case, the death of adult animals is rare. Feeds that are expected to be contaminated should be tested in a laboratory. Proper ventilation, dryness at the storeroom and use of toxin binder in feed prevent aflatoxicosis.

9. **Poisonous Plants:** It is important to note that several 100 plants are considered to be toxic to livestock under field conditions. Algae, Bracken fern, and nightshade are the most common poisonous plants.

Cattle will eat whatever they encounter if their stomach is empty, and eating a high proportion of toxic plant can have fatal or toxic effects. Fortunately, cattle that eat sufficient amounts of other feeds will rarely eat a high quantity of a poisonous plant to do any harm.

10. **Udder Edema:** Edema can be defined as an extreme gathering of fluid in the udder under the skin. This situation generally arises at calving and is more critical in first lactation cows. Prevention is by restrictive access to either potassium or sodium salts during the dry period.

It is also advisable to avoid any excessive grain. Treatment comprises of inspiring circulation by massaging the udder. Diuretics (drugs that endorse the creation of urine) should be used with utmost care and at the direction of a veterinarian.

11. **Urea Toxicity (Ammonia Toxicity):** It is usually seen that feeding of urea in large quantity at one time or inadequate carbohydrate consumption results in extreme ammonia (NH_3) in the rumen. It may cause animals to show symptoms such as show skin and muscle tremors, uneasiness, excessive salivation, incoordination, laboured breathing, and bloat. In addition, it may also cause animals to urinates unreasonably. No more than 0.2 kg of urea should be fed per cow per day. Providing vinegar, citric acid helps to decline the symptoms of urea toxicity.

6.8. CONCLUSION

At the end, it is concluded that dairy farm management is one of the most important tasks that play a critical role in the overall success of dairy. With the increase in dependency of people on a dairy farm for their daily needs, the burden on dairy farmers has enhanced. They can get success in the long run only if they achieve economies of scale. In addition, there are various things that they need to keep in mind as they are dealing with livestock products. Apart from this, they are required to be updated about the latest technologies so that to employ them in the farm to get good results. As tasks are complex, it is always recommended to use record keeping for various tasks. Such tasks will help them in ensuring sustainable practices that will help them in the long run.

REFERENCES

1. Rotz, C. A., Oenema, J., & van Keulen, H. (2006). WHOLE FARM MANAGEMENT TO REDUCE NUTRIENT LOSSES FROM DAIRY FARMS: A SIMULATION STUDY. *Applied Engineering in Agriculture, [online] 22(5), pp.773-784. Available at: <https://elibrary.asabe.org/abstract.asp?aid=21992> [Accessed 11 May 2021].*

2. (No author name found/given with ebook)SAI Platform Dairy Working Group,(2009).*Principles and Practices for Sustainable Dairy Farming*. [eBook] SAI Platform Dairy Working Group. Available at: https://saiplatform.org/uploads/Library/PPsDairy2009-2.pdf (accessed on 15 March 2021).

3. De Ondarza, M., & Tricarico, J., (2021).*Review: Advantages and Limitations of Dairy Efficiency Measures and the Effects of Nutrition and Feeding Management Interventions*. [eBook] Elsevier Inc. on behalf of the American Registry of Professional Animal Scientists. Available at: https://www.appliedanimalscience.org/article/S1080-7446(17)30086-4/pdf (accessed on 15 March 2021).

4. Dhaliwal, A., & Dhillon, G., (2017). Management practices followed by dairy farmers in rural and urban areas of Bathinda District in Punjab. *Journal of Krishi Vigyan, 6*(1), p.124.

5. Fao.org. (2021).*Dairy Farm Management*. [Online] Available at: http://www.fao.org/ag/againfo/themes/documents/PUB6/P612.htm (accessed on 15 March 2021).

6. http://www.fao.org/. (2021).*Guide to Good Dairy Farming Practice*. [Online] Available at: http://www.fao.org/3/a-ba0027e.pdf (accessed on 15 March 2021).

7. Israeli Dairy School Dairy Herd Management Training Center, (2021). *Seven Critical Principles in Dairy Herd Management-Israeli Dairy School Dairy Herd Management Training Center*. [Online] Available at: https://www.dairyschool.co.il/seven-key-principles-in-dairy-herd-management/ (accessed on 15 March 2021).

8. Kauffman, J., & Tauer, L., (1986). Successful dairy farm management strategies identified by stochastic dominance analyses of farm records.*Northeastern Journal of Agricultural and Resource Economics, 15*(2), 168–177.

9. Kildee, H., (2021).*Care, Feed and Management of the Dairy Herd*. [Online] Iowa State University Digital Repository. Available at: http://lib.dr.iastate.edu/iaes_circulars/15?utm_source=lib.

dr.iastate.edu%2Fiaes_circulars%2F15&utm_medium=PDF&utm_campaign=PDFCoverPages (accessed on 15 March 2021).

10. Milkproduction.com. (2018).*Guide to Good Dairy Farming Practice*. Milkproduction.com. [Online] Available at: http://www.milkproduction.com/Library/Editorial-articles/Guide-to-good-dairy-farming-practice/ (accessed on 15 March 2021).
11. Mwangi, D., (2008).*Dairy Cattle Management*. [eBook] Available at: https://elewa.org/animalmanagement/Dairycattlemanagement.pdf (accessed on 15 March 2021).
12. Patil, P.V. (2018). *Handbook of Animal Nutrition*. First edn. Write and Print publications, New Delhi, pp-245.
13. Penn State Extension, (2014).*Eight Things You Need to Know Before Starting Your Own Dairy Farm*. [Online] Available at: https://extension.psu.edu/8-things-you-need-to-know-before-starting-your-own-dairy-farm (accessed on 15 March 2021).
14. Pennington, J., (2018).*Dairy Production and Management Records*. [eBook] University of Arkansas, United States Department of Agriculture, and County Governments Cooperating. Available at: https://www.uaex.edu/publications/pdf/FSA-4005.pdf (accessed on 15 March 2021).
15. Priyadarshi, S., (2015).*Dairy Farm Management: Scope and Importance*. [Online] Biology discussion. Available at: https://www.biologydiscussion.com/essay/dairy-farm-management-scope-and-importance/5403 (accessed on 15 March 2021).
16. Publish.csiro.au. (2019).*Best Management Practices for Smallholder Dairy Farmers*. [Online] Available at: https://www.publish.csiro.au/ebook/chapter/9781486306473_Chapter_07 (accessed on 15 March 2021).
17. Putnam, L. D., & Karszes, J., (2021).*Dairy Farm Management*. [eBook] Department of Applied Economics and Management Cornell University Agricultural Experiment Station College of Agriculture and Life Sciences Cornell University, Ithaca, New York 14853-7801. Available at: https://ageconsearch.umn.edu/record/122646/files/Cornell_Dyson_rb0106.pdf (accessed on 15 March 2021).
18. Rao, P., (2014).*Dairy Farm Business Management*. Delhi: Biotech Books.

19. Snv.org. (2021).*Dairy Farm Management.* [Online] Available at: https://snv.org/cms/sites/default/files/explore/download/dairy_farm_management_training_manual_and_guideline.pdf (accessed on 15 March 2021).

20. Yadeta, W., Habte, D., Kassa, N., Befekadu, B., & Fetene, E., (2021). *Dairy Farm Record Keeping with Emphasis on its Importance, Methods, Types, and Status in Some Countries.* [Online] www.arcjournals.org/. Available at: https://www.arcjournals.org/pdfs/ijrsb/v8-i4/3.pdf (accessed on 15 March 2021).

Chapter 7

Importance of Cattle Breeding and their Characteristics

CONTENTS

In the chapter, the importance of cattle breeding and their characteristics, the basics of breeding management of cattle and buffaloes are discussed. Some of the factors that may impact the overall breeding efficiency are also discussed in this chapter. The chapter also shed some light on functional traits or characteristics. Management assessment with respect to nutrition and labour is also explained in this chapter. In addition, it also discusses crossbreeding systems. It also recognizes the importance of modified crossbreeding. Furthermore, genetic principles of cattle breeding, role of selection and culling are also explained. It also discusses some of the major performance traits and different types of mating systems in cattle breeding. At the end, crossbreeding for commercial beef production is also demonstrated.

7.1. INTRODUCTION

It is generally seen that the quality of cattle being produced by the beef industry is dependent on the genetic makeup of cattle and the management system to which they are subjected. Genetic makeup is generally under the total influence of breeders, both commercial and purebred, and this accountability should not be taken informally. Several researches show different types of cattle perform distinctly under diverse management conditions. It suggests that beef producers cannot just select for the greatest in traits of economic significance but that they should balance their genetics to their environment and resources. The selection of heifers and bulls, and the selected breeding system, influence the genetic class of the calf crop. The purebred producer generally makes use of selection methods for genetic management, while the commercial producer uses both mating and selection systems, particularly crossbreeding. As the majority of the sires are bought from them, purebred breeders carry a significant impact on the direction of the beef industry. Commercial producers are maintaining that the purebred seed stock producer sustains records and make these records readily obtainable. It is generally seen that both commercial and purebred producers recognize and use the tools and principles of genetic improvement.

7.2. BREEDING MANAGEMENT OF CATTLE AND BUFFALOES

Reproduction is generally considered an imperative consideration in the economics of cattle production. In the absence of regular calving and breeding at an adequate time, cattle rearing will not be lucrative. The ultimate objective of the dairy farmers is to produce a healthy calf every year. This can be done only by enhancing the reproductive efficiency of the animals.

Successful reproduction embraces the potential to mate, the capacity to nourish and to conceive the embryo and produce the feasible young ones at the close of a standard gestation period. In fact, disruption in this chain of events would result in the fiasco of the cow either to conceive or the embryo to die or to causing premature delivery of the fetus.

It is important to note that reproductive efficiency is a multifaceted phenomenon controlled by both non-genetic as well as genetic factors, the non-genetic factors being nutrition, climate, and level of management. The reproductive efficiency differs not only between breeds and species but also among the animals of the same breed.

Even the best management and feeding cannot persuade performance beyond the genetic limit of an inferior animal. Advancing the genetic merits of livestock populations are significant at each level of management. A sound breeding program is an essential element of the total animal production system.

7.3. BREEDING EFFICIENCY: FACTORS AFFECTING BREEDING EFFICIENCY

The factors that can impact the overall breeding efficiency of cattle are discussed in subsections.

7.3.1. Number of Ova

The first restraint on the breeding efficiency of the fertility of an animal is the sum of functional ova discharged during each cycle of ovulation. Ovulation can be defined as the process of shedding of the ovum from the Graffian follicle.

In the case of the cow, generally, a single ovum has the potential to undergo fertilization only for 5–10 hours. Hence, the time of mating insemination to ovulation is imperative for successful fertilization.

7.3.2. Percentage of Fertilization

The other restraint is the fertilization of ova. Unsuccessful fertilization may be because of several factors. The spermatozoa may be limited or extreme low in vitality. The service may be either too late or too early; so that the eggs and sperms do not congregate at the right time in order to facilitate the fertilization process.

7.3.3. Embryonic Death

It is important to note that from the time of fertilization till birth, there are a variety of reasons that may cause embryonic mortality. Hormone imbalance or deficiency may cause fiasco of embedding of fertilized ova which die successively. Death may happen because of lethal genes for which the embryos are homozygous. Some of the other factors may be overcrowding in the uterus, accidents in development, insufficient infections or nutrition in the tile uterus.

7.3.4. Age of First Pregnancy

There may be a significant decline in the breeding efficiency by enhancing the age of first breeding. Females which bred at a lesser age are more likely to look stunted during the first lactation, but their mature size slightly impacted by them having been bred early.

7.3.5. Frequency of Pregnancy

It is usually seen that the breeding efficiency can be significantly increased by reducing the interval between successive pregnancies. The overall standard policy is to breed for the first time at an initial age and to rebreed at almost the quickest opportunity after each pregnancy. In such a manner, the overall lifetime efficiency can be increased. Cows can be rebred in 9–12 weeks after parturition.

7.3.6. Longevity

The length of life of the parent is an essential element in breeding efficiency, as the return over feed cost is significant in the augmented length of life. In addition, it impacts the likelihood of improving the breed. The higher the life of the parents, the lesser the proportion of cows required for replacement every year.

7.4. CHARACTERISTICS OF CATTLE

The major functional traits important in the majority of the dairy farm include body size, age at puberty, milking potential, fleshing ability, hot climate adaptability, cutability, muscle expression, and marbling (Figure 7.1).

Figure 7.1. Characteristics of cow.

Source: *Image by www.reference.com.*

7.4.1. Body Size

Body size can be easily assessed as weight at a stated level of condition or body fatness. Though there are exemptions, weight at all ages tends to be positively linked—cattle which have comparatively higher weight are more likely to remain heavier throughout life.

It is often seen that genetically, animals that are larger in size gain weight faster, as their nutritional requirement is in accordance with their weight. A higher weight means a higher nutritional requirement.

Frame score, an objective measure of skeletal dimension used to guess current and future body size reveals that there is significant variation that happens in *Bos indicus* in body size relationships.

Calves from dams which are possessing *Bos indicus* inheritance is comparatively small at birth, generally developing in few problems in calving. Although, calves by *Bos indicus* sires out of *Bos taurus* dams generally are comparatively bigger in size, so calving may be more challenging.

7.4.2. Milking Potential

Milking potential refers to the genetic competence to produce milk. It is basically not the real volume of milk produced that is also subjective to the nutrition intake of cows as well as the calf's growth potential and nursing pressure.

To characterize milking potential accurately, it should be evaluated relative to body size. Higher-milking females need more nutrients for body maintenance and require higher-quality diets, even when not lactating.

7.4.3. Age at Puberty

Age at puberty relates to body size, milking potential and genetic classification. It is often seen that cows that are smaller in size and that give higher milk mature at an earlier, whereas *Bos indicus* mature comparatively later.

Although higher milking females, even the large ones, often reach puberty and conceive in their earlier age of life, subsequent fertility can suffer as their body may become thin after the starting of the lactation period. While *Bos indicus* types reach the age of puberty comparatively late, therefore, their productive life is generally longer.

7.4.4. Hot Climate/Tropical Adaptability

It is generally seen that some cattle have higher adaptability to hot climates, especially those whose parent is from hot climate regions as they carry inheritance. Animals which are lighter in colour have dark skin and short hair coats are generally best adapted to hot climate (Figure 7.2).

Figure 7.2. Hot climate/tropical adaptability of cow.

Source: *Image by Sustainability Times.*

High humidity strengthens the effects of heat, primarily because hot, humid climates usually add the stresses of low-quality forage and parasites. In areas where there is heat with humidity, cattle face the situation of significant stress as they fail to shed long, thick hair coats, especially the ones that have dark-coloured skins.

As might be expected, animals that have a high tolerance rate to hot climates are generally less adapted to cold climatic conditions.

7.4.5. Fleshing Ability

Fleshing ability can be defined as the capability of the body to fatten and retain fat. Fleshing ability believes to plunge with the surge in the size of the genetic body, maintenance requirements in comparison to size, inherent muscularity, and milking level. It is often seen that animals that are less adapted to the local environmental conditions are low in fleshing ability.

Animals that have heavy built such as *Bos indicus* often flesh more easily in comparison to other types on low-quality roughage and forage. Easy-fleshing cattle usually endure periods of nutritional energy deficiency more effortlessly and, therefore, can reproduce more steadily, but at the same time, they also over fatten more willingly in the feed yard unless they are managed properly.

7.4.6. Muscle Expression

Muscle expression is inherent muscularity, which means they are independent of other body tissues. Muscling is considered the second most crucial factor in cutability. Heavy-muscled types are generally low in fleshing ability; therefore, their reproductive efficiency is generally low.

7.4.7. Cutability

Cutability, or the percentage of lean, is generally assessed in beef carcasses as United States Department of Agriculture (USDA) Yield Grade. Cutability is often contingent on relative amounts of fat (that differs significantly), bone, and muscle (which varies least).

When genetic breeds or types are matched for variances in cutability, it is believed that the breeds have akin nutrition levels, but inherent cutability differences can be altered freely by unpredictable nutrition to attain comparable degrees of fatness.

7.4.8. Marbling

Marbling, or intramuscular fat, is one of the critical factors in defining USDA Quality Grade, an indicator of the palatability factors of juiciness, tenderness, and flavour. Marbling surges with age up to physiological maturity and usually is greater in cows that matures early and that give higher milk. Diet intake with high nutrients during the early period of life for extended periods usually increases marbling. Cows that are heavily built with heavy-muscled, low milking types have comparatively low marbling. Because marbling is associated with body fatness, especially in linking types or breeds, there is generally a trade-off between Quality Grade and Yield Grade. As one improves, the other ones tend to decline. One imperative consideration in ascertaining optimum functional levels is that categories of cattle may vary in comparative performance under unpredictable conditions.

7.5. MANAGEMENT ASSESSMENT

Management is one of the essential components that play a very important role in cattle breeding. To properly define the genetic type of cattle that is required, it is imperative to recognize what will be provided and how that influences the functioning of the herd. When evaluating management, the major areas of concern are nutrition availability and labour (Figure 7.3).

Figure 7.3. Management of cross-breeding dairy.

Source: *Image by Farmers Weekly.*

7.5.1. Nutrition

It is worth noticing that the accessibility and quality of nutrition are tremendously imperative when determining the breeding program. Nutrition plays a significant role as it influences how some cattle will perform under different conditions and how they are different from other cattle in performing different activities (Figure 7.4).

Figure 7.4. Nutrition assessment of cows.

Source: *Image by Bertoli.*

It was found in research that under nutritionally stressful situations, less-productive cattle that are smaller in size are more efficient at turning the resources present into pounds of saleable product. Their calves are generally smaller in size on average, but they believe to have better reproduction rates that compensate for the weight deficit.

Under ideal nutrition level, there were almost insignificant differences in efficiency between reasonably performing cattle and high-performing cattle. In an environment condition that provides more nutrition in comparison to what the cattle need the bigger, high performing cattle were the most effective at producing pounds of weaned calves or saleable product.

Based on this information, management operations that offer remarkable nutrition should cogitate more productive types of cattle; although, operations with poor nutrition, either in quality or availability, should consider less productive cattle (less milking ability and/or smaller in size).

Nutrition assessment should take into consideration forage base (infected fescue with high-legume mix/quality grass, sparse legumes, cool-/warm-season grass mix, etc.), the nutritional quality of stored feeds (hay harvested, silage, and stored correctly, hay harvested after optimal maturity and stored outside on the ground, etc.), and economical obtainability of purchased feedstuff s.

Quantity and quality of feed resources will play a significant role in taking management-related decisions, including breeding management.

7.5.2. Labour

Labour availability is also considered one of the important components when determining a breeding program. Even on farms that are entirely owned and operated by a family, labour is of significant consideration. It must be determined how attentively the cattle will be cared for.

In other ways, it also depends upon the type of farmers; for instance, if a farmer is working full time, then they can devote all their time, but if they have limited time, then they need external assistance.

Such sort of scenarios is very important as it helps in developing a breeding program. As an example, a full-time farmer who is keeping an eye on the cattle multiple times in a day may not have to pay as much attentiveness to get a calving-ease bull as the part-time farmer who seldom sees his cattle.

In addition, a full-time farmer also enables the opportunity to provide extra nutrition to the cattle when they are in need and can perhaps administer high-producing cattle efficiently in comparison to a part-time farmer who does not have adequate time to feed the cattle on a timely basis.

7.6. CROSSBREEDING SYSTEMS

It is important to note that crossbreeding systems should be planned for each operation, depending on the size of the herd, management level, potential market, and available facilities. A long-term plan is always essential to attain the maximum benefits from crossbreeding. Some advantages and disadvantages of various crossbreeding systems are discussed in subsections (Figure 7.5).

Figure 7.5. Crossbreeding in cows.

Source: *Image by The Feed-in A Drum.*

7.6.1. Two-Breed Terminal Cross

This system generally made use of straightbred cows and a bull of another breed. It would be a terminal cross if stopped at this point. One perfect example of this will be Angus cows bred to Charolais bulls.

In such a system, substitutions must be obtained from another source, or part of the herd (possibly young cows and heifers) would have to be bred to Angus bulls to make substitute heifers. It is usually not a wanted system because it does not recognize any heterosis in the cow as she is a straightbred.

7.6.2. Three-Breed Terminal Cross

This system uses a bull of a third breed and a two-breed cross (F1) cow. It yields the greatest hybrid vigour in the calf and cow. This is a terrific system because hybrid vigour is recognized for both maternal ability and growth rate. Replacement females for this system must be raised or purchased from another source. This is one of the best systems as it works for a herd of any size if there is the presence of high-quality replacement females.

7.6.3. Two-Breed Rotation or Crisscross

This is a naive crossbreeding system involving two breeding pastures and two breeds. A two-breed rotation is initiated by breeding bulls of breed B to cows of breed A. In each following generation, replacement heifers are bred to bulls of the breed which is the reverse of their sire.

Two breeds of bulls are needed after the initial 2 years of mating. The two breeds selected should be akin in accordance with birth weight, milk production, and mature size. It helps in lessening calving struggle in first-calf heifers and makes the management tasks easy:

1. **Three-Breed Rotation:** This system accompanies a similar pattern as the two-breed rotation, but a third breed is supplemented. The three-breed rotation upholds an advanced level of hybrid vigour in comparison to the two-breed system.

In this, mating plans can be mystifying, but individual cows are not conveyed from one breeding group to another. Three distinct groups of cows are ultimately produced, and they are mated to the sire breed to which they are minimally associated. This scheme endures for the entire life of a cow.

2. **Rotational-Terminal Sire Combination:** This system comprises the use of rotational mating of maternal breeds (breeds A as well

as breeds B) in a percentage of the herd to offer substitute females for the whole herd. The mature crossbred cows are then mated to the terminal sire breed (breed C).

All of the terminal cross offspring are marketed. It is usually seen that such a system upholds a greater level of production but also need great efforts of management. It almost does away with heifer selection as almost all heifer calves created by the rotational mating should be reserved to maintain adequate numbers of the herd.

3. **Heifers Out of Heifers:** This is one of the perfect examples of a rotational-terminal sire combination. There is no basis or reference for the argument that you should not keep a heifer out of a heifer. In contrast, this system is considered the finest one in attaining maximum efficiency.

In herds that have bull greater than one or where AI is likelihood, this is a productive crossbreeding system. Breed wholly heifers and sufficient younger females to total almost two-and-a-half times the sum of replacement females that are being planned to keep the next year to an easy-calving, virtuous maternal bull.

For instance, if 10 replacement females are anticipated the following year, breed 25 heifers and young females to the bull. Choice the entire replacement heifers out of this company of calves. Breed the remaining herd (mature cows) to a growth prosperous, great built terminal bull, and market all of the calves (refer to three-breed terminal cross).

Such a system permits the producer to get easy calving in the first-calf heifers and worthy maternal features in the replacement heifers and to amplify muscling and growth in the bulk of the feeder calves. Only downside is the nonconforming steers out of the heifers and young cows, but its advantages outweigh its downsides.

4. **AI Roto-Terminal:** This system generally makes use of a very rigorous synchronization program, and all heifers and cows are mated to a heifer-acceptable/maternal-type bull, using artificial insemination (AI). All cows (apart from the virgin heifers) are then exposed to a terminal type bull.

Virgin heifers that do not get pregnant by the first mating can be fertilized a second time, or otherwise, a greater number of replacements will be required to be preserved through pregnancy testing each year. It is important to note that heifers are only reserved from AI.

7.7. MODIFIED CROSSBREEDING

In the majority of the herds, the level of management and the facilities needed to use intricate crossbreeding systems are not present. Although, with some modification, one can make use of some basic crossbreeding principles. Here's how one can abridge the conventional systems: Purchase crossbred females. This is the easiest and speediest method of attaining extreme hybrid vigour (Figure 7.6).

Figure 7.6. Progress or change in cattle breeding.

Source: *Image by Beef Magazine.*

Bought two breed cross females can be bred to a terminal sire of a distinctive breed; this enhances both maternal and individual hybrid vigour. The producer requires an existing supply of disease-free and high-quality females. Use bull-breed rotation is basically concerned with taking into consideration a bull of one breed for a definite period of time (Suggestive time 4 years), then rotating to a different breed of bull.

If there is a need to maintain a balance between good replacement heifers and good feeder calves, swapping between breed types would also be needed. In other words, make use of a British breed bull for at least 4 years, then alter to a Continental breed for 4 years, then again change to the original breed. In addition, also put additional efforts to preserve a huge number of replacement heifers in years that a maternal-type bull is used.

There is only a need on one breeding pasture, and replacement heifers are produced within the herd. In this system, there are often sacrifices of few hybrid vigour when it is compared to a two-breed rotation, but it is straight forward in a manner that it is realistic for many producers.

7.8. GENETIC PRINCIPLES

To entirely understand breeding management, it is always essential to understand some basic genetic principles. Having a comprehensive understanding of the role genetics plays in each economically important trait of beef cattle can help in making prudent selection decisions efficiently and effectively.

It is imperative to have awareness about which traits can be transformed through breeding management (crossbreeding and/or selection) and which traits should be transformed by other management strategies.

The majority of the traits that are carrying economic importance (weaning weight, calving ease, etc.), in beef cattle are basically governed by two factors: the genetic makeup (genotype) of animals and the environment in which the animal lives.

The environment embraces not only the local environment conditions but also how the cattle are managed in such conditions. Creep feed, health programs, forage quantity, and quality are some examples of environmental effects.

Environmental effects on economically significant traits are generally governed through management techniques, such as health and nutrition programs, which will also be explained in other sections of this chapter.

The two sorts of genetic effects on economically significant traits of beef cattle are non-additive and additive. When a cow and bull mates, each one contributes almost 50% of its genetics to it's calf. If that calf is then permitted to reproduce, it conveys 50% of its genetics to each of its calves; although, each calf receives a distinctive sample of genes from its parents (because of this, generally sisters and brothers are distinct).

The sample of genetics that are generally received by offspring from their parents and ultimately passess on to their progeny is termed as the additive genetic effects. These are the genetic effects that are transferring from one generation to another; so, they are the centre of the selection. Heritability, which is the proportion of each trait governed by the additive genetic effects, is an imperative factor at the time of making selection decisions.

To comprehend it better, it is important to understand that offspring has additional similar characteristics as their parents for extremely heritable traits. In other words, the genetics that fosters parents to behave in a certain way would be passed on to the calves, and they would perform correspondingly. Highly heritable traits retort more quickly to the selection, while modest heritable traits reply more quickly to heterosis (crossbreeding) and management strategies (environment).

Another genetic effect that is significant while making selection decisions is genetic correlations. A genetic correlation happens when while selecting one trait, the other one is automatically affected. The effect of one trait on the other can either be disadvantageous or complementary.

An example of a complementary genetic correlationis: When selections are created for augmented weaning weight, yearling weight is also amplified. An example of a disadvantageous correlation is: as selections are made for augmented weaning weight, birth weight also surges.

The functioning of genetic correlations is almost the same, no matter what the trait is being selected for. In other words, as selections are taken to decline birth weights, yearling weights and weaning are generally plunged, too.

7.9. SELECTION AND CULLING

Selection and culling basically refer to the two sides of the same coin. Selection can be defined as a process in which some individuals in a population are involved in becoming the parents of the next generation. Instinctively, some are debarred for the objective to be culled (Figure 7.7).

Figure 7.7. Culling of cows from main herd.

Source: Image by Hoard Dairyman.

Natural selection is the most widely adopted method that has been going on for ages where stronger animals have higher survivability chances, and which were in more unison in relation to the local environment conditions, found a better chance to reproduce.

Thus, it will be right to say that certain genes for firm characters got more chance to be chosen to reproduce individuals in the following generations. Since the domestication of cattle, people are in search of superior phenotypes in traits that are beneficial to them and thus choosing such animals to form the parental generation.

It can be called man-made artificial selection. Now scientists pertinent in this field have advanced one step further in making estimates of genotypes from the study of phenotypes and making use of that information (in artificial) selection.

- **Selection Methods:** In the selection method, there is only one and the best way that is to keep the best and cull the poorest. The several selection methods are techniques that can be used in estimating or identifying the genetic values of individual candidates for selection. The procedure that is discussed below applies to the selection of quantitative trails.

7.9.1. Performance Testing

Performance test can be defined as estimating the phenotypic value of the individual candidates for selection. Since the phenotypic value is discovered by both environmental and genetic influences, the performance test is just an estimate, not a computation of the genetic value. The happening of this approximation relies upon the heritability of the trait, for instance; to the extent up to which the genetic value is modified by the environmental influences.

- Advantages:
    - Among simple procedures, the performance test is considered as one of the most accurate ones;
    - Environmental influences can be curtailed by testing candidates for selection in similar environmental conditions or the same pen;
    - The measure is direct, not on a comparative basis;
    - In this, all the folks can be tested before selection in contrast

to the progeny testing in which only parents test can be done;

- In this type of testing, generation intervals are mostly of the shorter period;
- Testing generally done at the farm under normal management conditions.

- Disadvantages:
    - In this type of testing, the level of precision becomes low when heritability is low;
    - There is no availability of phenotypes for one sex or in sex-limited traits such as milk yield;
    - Traits that are not articulated until maturity may become problematic or costly to manage by performance tests as the majority of selection related decisions should be done before maturity.

Performance tests are often considered the backbone of most selection programs. Although others method is also quite popular as they easily get more publicity, it remains a fact that majority of the development done in livestock improvement to date has been because of selection on the individual's own phenotype, i.e., performance test.

7.9.2. Pedigree Selection

A pedigree can be defined as a record of an individual's ancestors, including its parents. This information is very critical as each individual owns a sample of half of the genes from each parent. If scientists can exactly know an individual's phenotype, little can be achieved by considering pedigree in selection.

Pedigree plays a very important role when we do not possess adequate correct records with respect to the production of the individual. Also, it is very important in the early selection when the traits in question might not have expressed themselves. It is also very significant for selecting males when the traits selected for are articulated only by the female, such as milk production in dairy cattle.

- Advantages:
    - It provides information when the availability of performance tests is very low for candidates;

- It provides information that can be useful in supplementing performance test information;
- It allows selection to be complete at a younger age. Pedigree records would be useful in selecting animals for performance or progeny testing in a multi-stage selection scheme;
- In this, the selection of bulls can be certain on the basis of milk records of their female relatives.

- Disadvantages:
- In this type, accuracy is very less in comparison to other selection procedures;
- Too much emphasis on relatives, particularly remote relatives, significantly lessens genetic progress;
- Progeny of preferred parents are generally environmentally favoured;
- Relatives usually make records under quite distinctive environments, thus presenting non-random bases into the selection system.

7.9.3. Progeny Testing

In this method type, the breeding value is determined by engaging in a study of the expression of the trait in its offspring. Individuality helps in assessing what an animal seems to be, his pedigree expresses what he should be, but the functioning of his progeny states to us what actually he is.

Progeny testing is basically a two-stage selection system as certain initial selection defines which animals produce progeny firstly accompanied by additional culling of these that generate poor progeny.

- Advantages:
  - It usually carried higher accuracy when many progenies are attained.
- Disadvantages:
  - It usually has a long generation interval period;
  - It generally needs high reproductive rate;
  - It usually has low selection intensity.

7.9.4. Show Ring Selection

Selection based upon show ring performance has had significant importance in the past. Basically, this selection has been aimed at taking the conformation of the animal to some perfect conformation.

This development has been centred on two basic objectives:

i. Improvement conformation; and

ii. Correlated response.

Improvement of conformation has an economic value as a part of the sale price is decided by the conformation of the individual. The perfect sort was chosen so that, in the judgment of the experts, the animal possessing this conformation was most probable to be a lucrative producer. In other terms, the judges were aimed at stressing traits of conformation that are interlinked with productive ability. With the introduction of record-keeping, it was witnessed that direct selection for performance traits stemmed in much rapid development than selection through correlated conformation traits. Also, when exposed to intensive study, many of the associations between performance and show ring were discovered to be of non-genetic origin. In case the relationships are of genetic origin, direct selection for performance should enhance the reverse situation as well as the confirmation. The show ring is considered as the best platform for discussion of what composes idyllic type and worthy management and has fabricated significant alterations in the conformation of some species. This has arisen chiefly from the education of the breeders, though, for the majority of the animals that are existing in the ring are good and selection difference among these animals is typically so meagre as to harvest little change.

- Advantages of Show Ring Selection:
  - It fosters breeders to exchange experience as well as ideas;
  - It made the comparisons possible among animals of superior quality both between and within breeds;
  - It allows new breeders to establish connections with established breeders.
- Disadvantages of Show Ring Selection:
- In this selection method, the main focus is on traits that carries insignificant economic importance;
  - Showmanship and clever fitting can conceal flaws of several sorts;

- It is generally seen that in this method type, differences between exhibited animals are generally trivial;
- Production and conformation traits typically have low genetic correlations.

7.10. MAJOR PERFORMANCE TRAITS

It is worth noticing that all the traits of economic value should be taken into consideration while selecting beef cattle (Figure 7.8). The major traits that can impact the overall production efficiency of desirable beef are discussed as follows:

- Reproductive performance or fertility;
- Maternal ability;
- Growth rate;
- Feed efficiency;
- Body measurements;
- Longevity;
- Carcass merit.

Figure 7.8. Performance traits in cattle.

Source: *Image by ABS global.*

It is often seen that maximum production efficiency is not always related to maximum performance levels in all of these traits because of adverse genetic relations between and within certain traits.

For instance, the large size of cows and high production levels of milk are correlated with a speedy growth rate in the calf but are not wanted when there is a limited feed supply because reproduction in the cow is significantly impacted.

7.10.1. Fertility

Good quality of reproductive performance or fertility is central to an effective beef cattle enterprise. Fertility is generally considered as a common measure with respect to the calf crop percentage, and no solo element in commercial cow-calf operations has a larger bearing on overall production efficiency in comparison to the number of calves weaned per cow in the herd. The percentage of calf crop can vary from 70% to 95%. However, the heritability of calving interval or fertility is low, which is just 10%.

Therefore, the majority of the deviation in calving proportion is because of environmental factors such as management, feeding, or herd health.

It is well known that fertility is a complex trait. Many genetic and environmental and factors impact the overall fertility from the time a cow is turned into a bull until her calf is weaned.

Basic cow herd records should report calf crop proportion in order to evaluate if the problem still exists.

Where calf crop or fertility fraction is low, there is a need for comprehensive reports records on reproductive traits in order to determine nutrition, management, genetic or herd health problems that can be altered to enhance reproductive performance in the herd. The connotation between fertility and other performance traits, both negative as well as positive, should be taken into consideration.

For instance, selection for heavy weaning weight may result in higher production of milk or bugger size of a cow in the herd or both and require a superior nutritional level for the cow herd to ensure an adequate fertility level.

While on the other side, culling open cows or problem breeders to enhance the reproductive performance in the herd with no favour for the concerned reason for low reproductive performance will reduce the overall milk production capacity and cow size in the herd.

Also, a higher weight at the time of birth is concerned with an increased rate of gain and mature size. Calving trouble as a result of hefty birth weights can cut deeply into calf crop weaned by decreasing calf survival at birth and commencement rate in the cow the succeeding breeding season.

7.10.2. Maternal Ability

The potential of a cow to wean a healthy, energetic calf is essential to effective beef production. Increased milk production surges weaning weight per calf, and heavier weaning weights can enhance the overall production efficiency with respect to the total fixed costs for all herd. Although, the requirement of feed and cost of feed is dependent on the size of the cow and its milk production capacity.

Therefore, milk production should comply with the feed resources to enhance the overall production efficiency. Optimum milk production is neither minimum nor maximum milk production in several circumstances.

Increased weaning weight per calf from the production of milk can be negative if weaning weight per cow is abridged because of poor rebreeding performance or market value is lessened by a wasty condition.

7.10.3. Growth Rate

The growth rate plays a very important role due to its significant correlation with the economy of gain with respect to fixed costs. The growth rate has generally been evaluated in time constant, post-weaning feeding tests.

An equitable high level of feeding for at least 112 days is needed to assess differentiation in growth rate most precisely. Genetic correlations among measures of size or growth at distinctive ages are generally high. Selection for the rapid rate of gain in post-weaning feeding tests generally enhances both mature size and weight at the time of birth.

It is often seen that with the increase in weight, there would be greater difficulty in calving. Increased mature size reduces the quality of carcass when slaughtered at standard market weights and escalates the nutrient necessities for the upkeep of the cowherd.

Preferably, the beef animal should have reasonable weight during the time of birth, and grow gradually, but mature and finish for slaughter early. But, the selection criterion to attain such a growth curve is intricate.

7.10.4. Feed Efficiency

Feed efficiency refers to a trait considered of significant economic importance in beef cattle. It is almost challenging to assess feed efficiency because individual feeding and adjustments for alterations in weight are needed.

Increased weight is concerned with greater feed requirements per unit of gain. To get more insights, it is highly important to measure feed efficiency in feeding tests devised within the framework of current-day cattle marketing and feeding practices.

It is often seen that breeders primarily relied upon alterations in the rate of gain as a display of feed efficiency instead of incurring the added expense of individual feeding. Some bull testing stations attain distinct feed consumption information to gauge the pounds of feed needed per pound of animal gain.

7.10.5. Body Measurements

Objective body measurements are very helpful in selection aids. Some of the shared measurements of cattle include height at the shoulder, backfat, height at the hips, depth of body, length of the body, pelvic size and scrotal circumference.

Linear body measurements play a significant role in comparing mature animal size to production resources. Body measurements were never planned for the objective to select for a larger size. These dimensions should also never be implicit as a substitution for the weight of a beef animal at a specific age. No particular size for an animal will be perfect for all management and feed resources, feed costs and breeding systems. Market weight and reproductive rate eventually conclude the optimal range in size for given set management and of feed resources, production costs, and breeding systems.

Hip height measurement is one of the most widely used body measurement in selection programs. The suggested point for measurement of height is straightforwardly over the hooks or hip bones. These measurements may be transformed to frame scores using the heifer hip and bull and height tables in the Appendix.

7.10.6. Longevity

It is important to note that the longer the animals remain productive in a herd; the fewer need will be required for replacement. Thus, the price of growing replacements to productive age is abridged. Longevity is extremely crucial in the commercial cattle herd.

Some of the factors that may impact the longevity of cows are unsoundness of feet, infertility, and legs, unsound mouth and udder troubles.

Purebred breeders are basically interested in making genetic improvement in longevity for the advantage of the commercial beef cattle population. Although, animals held in purebred herds to an old age surge the generation interval, which lessens the probable rate of genetic improvement from the selection. Selection for longevity must be restrained chiefly to such indicators as structural soundness.

7.10.7. Carcass Merit

Carcass merit also carries some significance in the beef cattle industry. The attractiveness of the beef product governs the price that consumers are ready to pay and the price at which they are ready to purchase a particular product.

Consumers usually demand a beef that has a high proportion of lean in comparison to bone and fat, and the lean must be flavorful, tender, and juicy. The variance in the proportion of lean, generally known as cutability, is the main factor inspiring distinction in carcass value. It is common for carcasses of the same quality to array from 10 to 30% fat trim. Although, such differences in composition are because of both genetic and environmental variations.

Beef quality that embraces such characteristics as flavour, tenderness, and juiciness, can have an imperative impact on consumer beliefs, value, and acceptance. Beef quality is assessed by texture, marbling, colour, maturity, and firmness.

The genetic connotation between quality and cutability is negative. Selection for high quality will generally curtail cutability, and selection for high cutability will generally diminish the quality of beef. In addition, selection for mature size or rate of gain often will lessen the quality of the carcass.

7.11. MATING SYSTEMS

The five fundamental types of mating systems are:

1. **Random Mating:** It refers to mating individuals without regard to the likeness of pedigree or likeness of performance.
2. **Inbreeding:** It is mating individuals that are more closely related than the average of the population or breed. Line breeding is

considered as a specific type of inbreeding and signifies the mating of individuals, so the correlation to a certain individual is either increased or maintained. Line breeding results in some inbreeding as associated individuals should be mated. Close breeding and Line breeding are the two types of inbreeding. Close inbreeding means mating between sire to daughters or son to dam and line breeding means mating between half-brother and sister or mating of animals more distantly related.

3. **Outbreeding:** It is basically referring to the mating of unrelated individuals Outcrossing, cross-breeding, Species hybridization/heterosis and grading up are the four types of out breeding. Outcrossing is the mating of unrelated pure bred animals within the same breed. Whereas, the mating of animals of different breeds is called crossbreeding. Heterosis or hybrid vigour is a phenomenon in which the crosses of unrelated individuals often result in progeny with increased vigour much above their parents. Grading up is the practice of breeding sires of a given breed to non-descript females and their offspring for generation after generation.
4. **Assortative Mating:** It refers to the mating of individuals that are more similar or akin in performance traits in comparison to the average of the group or herd. Such sort of mating system is generally made into use when there is a need to establish uniformity within a herd.
5. **Disassortative Mating:** It refers to the mating of individuals that are less similar/akin in performance traits in comparison to the average of the group or herd. This mating system is used to precise deficits within a herd.

7.12. CROSSBREEDING FOR COMMERCIAL BEEF PRODUCTION

Crossbreeding is basically used in commercial beef production to realize hybrid vigour, heterosis, and complementary combinations of breed characteristics and to match feed, market demands, and other resources present in specific herds. The effect of heterosis on some performance traits is also valuable. Productivity in some traits is better in crossbred animals than the average of the two parents because of heterosis. The heterosis effect is negatively proportionate to heritability. For instance, high heritability traits like feed efficiency, post-weaning growth rates, and carcass composition are impacted less by heterosis than low heritability traits such as fertility and liveability.

More than 50 breeds are present in substantial numbers to cattle producers either through AI or natural breeding. These breeds vary significantly in performance traits. Because offspring that are the outcome of crossbreeding tend to be a combination of both parents, crossbreeding can be made into use to access a performance goal in 1 or 2 years that would need numerous years to attain through selection for genetic change within one breed.

In addition, by matching the dam breed and the sire breed in appropriate combination, complementary traits can be attained in the offspring through the process of cross breeding. For instance, a cow chosen for its quick maturity, small size, low maintenance cost and high fertility can be coordinated with a sire breed selected for speedier growth rate and a lean muscular carcass.

7.13. CONCLUSION

There are basically two imperative practices that commercial producers should take into consideration in their beef breeding program: crossbreeding and selection. Breeds should be chosen based on their potential to attain economic/production goals and their capability to fit the production environment (management).

These breeds would be useful in a planned crossbreeding program that sustains a high level of heterosis. Once the breeds are identified, the selection of individual bulls would be easy on the basis of their performance level and the one that is a desire for each trait of economic importance. Overall, the performance of the breed for these traits should be taken into notice while determining this level.

Bulls should also be assessed or evaluated visually and be sound breeders. If such steps are followed suggestively, one can easily customize their cow herd to attain their breeding goals within the prevailing production environment.

REFERENCES

1. Acs.edu.au. (2021).*Breeding Dairy Cattle*. [Online] Available at: https://www.acs.edu.au/info/livestock/cattle/breeding-dairy-cattle.aspx (accessed on 15 March 2021).
2. Agritech.tnau.ac.in. (2018).*Breeding Management of Cattle and Buffaloes*. [Online] Available at: http://www.agritech.tnau.ac.in/expert_system/cattlebuffalo/Breeding%20management%20of%20cattle%20and%20buffaloes-2.html (accessed on 15 March 2021).
3. Agritech.tnau.ac.in. (2021).*Selection of Dairy Cattle*. [Online] Available at: https://agritech.tnau.ac.in/animal_husbandry/animhus_cattle%20-%20selection.html (accessed on 15 March 2021).
4. Appohigh.org. (2021).*Breeds of Beef Cattle*. [Online] Available at: https://www.appohigh.org/ourpages/auto/2010/12/8/52590465/breedsofbeef.pdf (accessed on 15 March 2021).
5. Banerjee, G.C. (1999). *A textbook of Animal Husbandry*. 8th edn., Oxford & IBH publishing Co. Pvt. Ltd., New Delhi, India, pp-1079.
6. Breeding and Genetic Selection, (2018).*Breeding & Genetic Selection*. [Online] Available at: https://www.uaex.edu/farm-ranch/animals-forages/beef-cattle/breeding-genetic-selection.aspx (accessed on 15 March 2021).
7. Bullock, D., (2018).*Planning the Breeding Program*. [Online] Www2.ca.uky.edu. Available at: http://www2.ca.uky.edu/agcomm/pubs/id/id108/05.pdf (accessed on 15 March 2021).
8. Cattle Reproduction Animal Husbandry: Home. [Online] Available at: http://14.139.187.9/animal_husbandry/animhus_cattle_reproduction.html (accessed on 15 March 2021).
9. CCAFS: CGIAR research program on Climate Change, Agriculture and Food Security, (2021).*Animal Breeding Benefits Farmers, Offers Food Secure Opportunity for Mitigating Climate Change*. [Online] Available at: https://ccafs.cgiar.org/news/animal-breeding-benefits-farmers-offers-food-secure-opportunity-mitigating-climate-change (accessed on 15 March 2021).
10. Freeman, A. E., (2021).*Animal Breeding-Immunogenetics*. [Online] Encyclopedia Britannica. Available at: https://www.britannica.com/science/animal-breeding/Immunogenetics (accessed on 15 March 2021).

11. Gadberry, S., Jennings, J., Ward, H., Beck, P., Kutz, B., & Troxel, T., (2019).*Beef Cattle Production*. [eBook] University of Arkansas, United States Department of Agriculture, and County Governments Cooperating. Available at: https://www.uaex.edu/publications/pdf/mp184/MP184.pdf (accessed on 15 March 2021).
12. Gowane, G., Kumar, A., & Nimbkar, C., (2019). Challenges and opportunities to livestock breeding programs in India.*Journal of Animal Breeding and Genetics, 136*(5), 329–338.
13. Hammack, S. P., (2021).*Texas Adapted Genetic Strategies for Beef Cattle V: Type and Breed Characteristics and Uses*. [eBook] The Texas A&M System. Available at: https://animalscience.tamu.edu/wp-content/uploads/sites/14/2012/04/geneticsE190.pdf (accessed on 15 March 2021).
14. Henderson, A., (1977).*Cattle Breeding*. Rome: Food and Agriculture Organization of the United Nations.
15. http://www.fao.org/. (2021).*Selection Programs*. [Online] Available at: http://www.fao.org/3/x6536e/X6536E04.htm (accessed on 15 March 2021).
16. Miglior, F., Loker, S., & Shanks, R., (2013). Dairy cattle breeding. *Sustainable Food Production*, 740–746.
17. MLA Corporate, (2021).*Breeding Objectives and* Selection. Meat & Livestock Australia. [Online] Available at: https://www.mla.com.au/research-and-development/Genetics-and-breeding/breeding-objectives-and-selection/# (accessed on 15 March 2021).
18. Mugisha, A., Kayiizi, V., Owiny, D., & Mburu, J., (2014). Breeding services and the factors influencing their use on smallholder dairy farms in Central Uganda.*Veterinary Medicine International,2014*, 1–6.
19. Peda.net. (2021).*Livestock Selection and Breeding*. [Online] Available at: https://peda.net/kenya/css/subjects/agriculture/form-3/lsab (a ccessed on 15 March 2021).
20. Wiki.groenkennisnet.nl. (2021).*Chapter 1.1 The History of Animal Breeding: Science and Application-Animal Breeding-Groen Kennisnet Wiki*. [Online] Available at: https://wiki.groenkennisnet.nl/display/TABChapter+1.1+The+history+of+animal+breeding%3A+science+and+application (accessed on 15 March 2021).

Chapter 8

Current and Future of Animal Nutrition

CONTENTS

Animal nutrition plays a very important role in the overall functioning and growth of animals. In the chapter, current and future of animal nutrition, the basic meaning of animal nutrition is explained. The chapter also explains the nutrient requirements for several types of livestock.

In addition, the chapter also shed some light on the formulation of least cost rations. It also discusses some biotechnological and allied applications employed in livestock nutrition. It also highlighted early developments in animal feed production. It also demonstrates the U.S. feed industry. Moreover, it also tries to explain the future of feed. At the end, the relationship between the information age and the feed industry is discussed.

8.1. INTRODUCTION

Nutrition can act as a major roadblock in the production of livestock, particularly when feed resources are insufficient in terms of both quantity and quality. Over the period, global production of livestock has increased dramatically, resulting in an increased population of the animal. However, these resultant increases in the animal populations have not always been supplemented with enhanced availability of livestock feed resources.

These may result in erosion, overgrazing, performance, and reduced health. Feed quality and quantity mingled with low producer costs have often enforced feed producers and farmers to remain at low levels of animal feed production, balanced by large numbers of animals.

It is well known that high global population growth complemented by high future demand forecasts for livestock products; emphasize the demand for higher yield per animal as well as surges in the number of animals. Insufficient quantity and feed quality hinders increased production of animal.

As it is expected that the world population is likely to increase from 6 billion to around 8.3 billion by 2030 at an average compound annual growth rate (CAGR) of 1.1% per year, it is imperative to be prepared in advance to produce adequate food for the increased population-based on local availability of feed resources, particularly in the developing countries.

It is widely believed that there are both opportunities and challenges for researchers to increase the productivity of animals with respect to both quality and quantity, through the application of suitable technologies in nutrition, production systems, and feeding of livestock.

Feed is basically considered as the most significant input in all livestock production systems in context to the cost, and the accessibility of economical, high-quality feeds is vital if production of livestock would remain competitive and continue to flourish to fulfil the demand for animal protein.

A researcher stated that traditional methods of livestock improvements (livestock nutrition, genetics, and breeding, and livestock disease controlling) have been used in the past and fulfilled the objective of enhancing livestock productivity.

Although, these methods are not sustainable for a longer period; subsequently, new intensive techniques such as biotechnology are the need of the hour to augment productivity. Modern biotechnology possesses the capability to provide new prospects for attaining higher livestock productivity in a way that lessens poverty, enhance food security and nutrition, and endorse sustainable consumption of natural resources.

Considerable improvement has been witnessed in livestock nutrition and feeding over the past two decades. Worldwide, the production of livestock has been increasing at a rapid rate in comparison to another sector, and it is estimated that by the year 2020, livestock would likely become the most significant agricultural sector with respect to the added value.

In research conducted, it was also stated that the feeding of genetically engineered (GE) crops to livestock for the last 15 years has exhibited compositional uniformity and similar levels of safety between GE crops and their traditional counterparts.

It was also stated that present production demands on the livestock industry have been concentrated against the use of antibiotics as growth supporters because of mounting concern over microbial antibiotic resistance.

Thus, with many countries witnessing higher incidences of antibiotic-resistant bacteria, regulations, and laws should be reformed to end in-feed antibiotic use in the animal production industry. This necessitates the need for adequate alternatives to be established for the addition of livestock feed.

Several reports have revealed proof that permitted nanoparticles and probiotics may be better substitutes for animal growth promotion and antimicrobials. However, it was believed that despite the growth of antibiotic resistance in bacteria, antibiotics have yet not been rendered entirely unsuccessful against them.

And that efficacy and the delivery of antibiotics could, though, be augmented by nanoparticle carriers, thereby possibly lessening the dosage of antibiotics needed for treatment. Recent advancement in the nutrition of livestock, particularly in monogastric, have concentrated on three main aspects:

- Developing the thoughtful of nutrient necessities of livestock;
- Determining the availability and supply of nutrients in feed ingredients; and
- Devising least-cost diets that help in fulfilling nutrient demand and supply together effectively and efficiently.

8.2. WHAT IS ANIMAL NUTRITION?

Animal nutrition can be described as ensuring or fulfilling the requirement of nutrition in livestock in order to maintain acceptable performance of neonatal, finishing, growing, and breeding animals (Figure 8.1).

Figure 8.1. Zinc oxide animal nutrition.

Source: *Image by EverZinc.*

From a practical standpoint, an adequate nutrition diet should ensure enough intakes of amino acids (both conventionally categorized essential as well as nonessential), fatty acids, carbohydrates, vitamins, and minerals, by animals through a supplementation schedule that fixes deficiencies in basal diets if there is any (e.g., soybean, and corn-based meal diets for swine; milk replacers for lambs and calves; and existing forage for ruminants).

8.2.1. Benefits of Dietary Supplements

Additionally, dietary supplementation with definite nutrients (e.g., glutamine, arginine, conjugated linoleic acid, and zinc) can normalize key metabolic pathways and gene expression to enhance fertility, immune function, pregnancy outcome, neonatal growth and survival, meat quality, and feed efficiency.

Overall, the proper balance of energy, protein, vitamins, and all nutritionally essential minerals in diets is required to make an efficient nutrition diet that is both economical and productive. Both fundamental research and practical application required to fulfil this goal.

8.2.2. Reaching Adequate Water Intake

Apart from this essential thing for an effective nutrition program, water intake is equally important. Livestock may encounter health issues if the quality of water offer to them is not up to the needed standard. It is important to note that consumption of water is far more important than consuming food. A successful livestock farm management requires an adequate and good standard water supply, in terms both of quality and quantity.

Safe and consistent supplies of water are extremely essential for livestock. If livestock does not take the adequate intake of safe water every day, intake of feed (concentrates and roughages) will drop, production will plunge, resulting in loss of money to the owner of a dairy farm.

8.3. NUTRIENT REQUIREMENTS FOR LIVESTOCK

Nutrient requirement basically refers to ensuring the minimum levels of nutrients for distinctive livestock species. It is important to note that livestock should be fed differently following their body type as well as the age, type of species, and purpose of production (Table 8.1).

Table 8.1. Nutrients Requirements for Livestock

Laying Chickens	Nutrient requirements for laying chickens consuming between 80 and 120 g/hen/day are as follows: 2.71–4.06% calcium, 12.50–18.80% crude protein, 0.13–0.19% mg/kg potassium, 0.21–0.31% non-phytate phosphorus, 0.13–0.19% sodium, and 29.00–44.00 mg/kg zinc.
Broiler Chickens	Broilers of age between 0 and 8 weeks old needs the degree of nutrients as follows: 0.80–1.00% calcium; 18–23% crude protein; 0.30–0.45% nonphytate phosphorus; 8.00 mg/kg copper; 0.30% potassium; 0.123–0.20% sodium, and 40.00 mg/kg zinc.
Broiler Breeders	Broiler breeders needs the following ranges of nutrients: 19.5 g/day crude protein, 4.0 g/day calcium, 350.0 mg/day nonphytate phosphorus, and 150 mg/day sodium

Turkeys (12–24 weeks old)	Turkeys 12–24 weeks old require the following ranges of nutrients: 14.00–19.00% crude protein, 0.55–0.75% calcium, 0.28–0.38% nonphytate phosphorus, 6.00 mg/kg copper, 40.00 mg/kg zinc, and 0.12% sodium.
Turkey Poults	Turkey poults at 0–12 weeks old require the following ranges of nutrients: 22.0–28.0% crude protein, 0.85–1.20% calcium, 0.42–0.60% nonphytate phosphorus, 6.00–8.00 mg/kg copper, 50.00–70.00 mg/kg zinc, and 0.12–0.17% sodium.
Turkey Tom Breeders	Turkey tom breeders require the following ranges of nutrients: 12.00% crude protein, 0.50% calcium, 0.25% nonphytate phosphorus, 6.00 mg/kg copper, 40.00 mg/kg zinc, and 0.12% sodium.
Turkey Hen Breeders	Turkey hen breeders require the following ranges of nutrients: 14.00% crude protein, 0.25% calcium, 0.35% nonphytate phosphorus, 8.00 mg/kg copper, 65.00 mg/kg zinc, and 0.12% sodium

Source: *Table by ResearchGate.*

The suggestions only serve as guidelines used for selecting dietary nutrient (essential amino acids, protein, crucial fatty acids, vitamins, and minerals) concentrations in practical diets. Most nutrients are attained from the digestion of feedstuffs, but few such as vitamins, minerals, and some critical amino acids are generally conveyed as synthetic supplements, especially in monogastric.

8.3.1. Formulation of Diets for Poultry

Poultry that is raised under proper guidance should be fed an appropriately diet on the basis of their age, species, and purpose of production. Table 8.1 provides a summary of recommended minimum levels of selected nutrients for layers, broilers, broiler breeders, turkey poults, turkey growers, turkey tom breeders, and turkey hen breeders.

In poultry, particularly in chickens, since each specific genotype has its own requirements, most commercial feed formulations are carried out based on minimum requirements recommended by the breeding companies from which they were obtained.

8.3.2. Formulation of Diets for Pigs

Several feed ingredients deliver nutrients that are needed by pigs for their routine or day to day performance. It is important to note that pigs do not need specific ingredients in their diets, but rather needed nutrients and energy such as minerals, amino acids, and vitamins.

The diets given to them should be balanced in context to amino acids, comprising passable levels and ratios of the 10 essential amino acids needed by pigs for growth, maintenance, lactation, and reproduction.

The 10 essential amino acids are histidine, arginine, isoleucine, lysine, leucine, methionine, threonine, phenylalanine, valine, and tryptophan. It is worth considering that in pigs; amino acids are described to be the chemical components of protein and are usually conveyed to the pig from the crude protein in the diet.

Failure to complement a low protein diet or feedstuff with adequate amounts of fine quality protein source was witnesses in many farms, which results in insufficient feed utilization, poor growth, general thriftiness, augmented carcass fatness, and or low reproductive performance.

It is also important to note that in pigs, dietary crude fibre should not surpass more than 10–15% of the diet as feed intake may be down. Growing and lactating pigs should be fed *ad libitum* while others could be limitedly fed. Presented in Table 8.2 are some amino acids needed for the growth of pigs.

Table 8.2. Amino Acid (%) Requirements for Pigs

Amino Acid	Growers	Pregnancy	Lactation
Arginine	nd	0.15	0.41
Histidine	nd	nd	0.37
Isoleucine	nd	0.42	0.46
Lysine	1.10	0.43	0.55
Methionine	0.26	0.12	0.30–0.36
Methionine/ cystine	0.57	0.06	nd
Phenylalanine	nd	nd	nd
Threonine	0.60–0.70	0.41	0.42
Tryptophan	0.18–0.20	nd	0.12
Valine	nd	0.32	0.53–0.68

Source: *Table by ResearchGate.*

8.3.3. Formulation of Diets for Fish

Fish farmers are often required to consider well-balanced, less pricy feeds as well as effective fish farming management practices to make production profitable. Species-specific feed preparations, which fulfil the nutritional need of fish at different stages of life, are required in fish farming. In addition, each exact genotype has its own nutrient necessities that fulfil the necessity at different stages of life.

The production of fish larvae and nutrition are typically commenced by breeding companies that are experienced in this field.

Most commercial fish diets or feeds are prepared on the basis of minimal necessities suggested by the breeding companies that convey the fingerlings or fry. Fish require nutrients such as crude protein, essential fatty acids, vital amino acids, lipids, crude fibre, minerals, carbohydrates, and vitamins.

1. **Ingredient Composition for Different Life Stages of Fish:** Some of the ingredients that are needed in early fry to brooder stages are as follows: fish meal of between corn meal of 0–4%, 30 and 68%, poultry by-product meal of 2–8%, fish oil of 9–12%, ground wheat of 17–22%, a mineral premix of 0.5% and vitamin premix of 1.5%.
2. **Feed Parameters and Proximate Composition for Different Life Stages of Fish:** Nutrient requirements between early fry and brooder stages are as follows: up to 6 months shelf life of feed, 3–8% of body weight, supplement of probiotics to enhance the efficiency of feed conversion, 35–48% crude protein, 8–21% crude lipid, 2–5 mm pellet size (mash for early fry), 17–21 kJ/g gross energy, and 12–13% nitrogen-free extract, 9–12% ash, less than 3–6% crude fibre.

8.3.4. The Feeding of Ruminants: Cattle, Sheep, and Goats

It is important to note that ruminants have a definite benefit over monogastric as it possesses the ability to alter organic materials that are not appropriate for human consumption into products having a high nutritional value such as milk, meat, and by-products. They also provide fertilizer from the undigested and faecal residues (Figure 8.2).

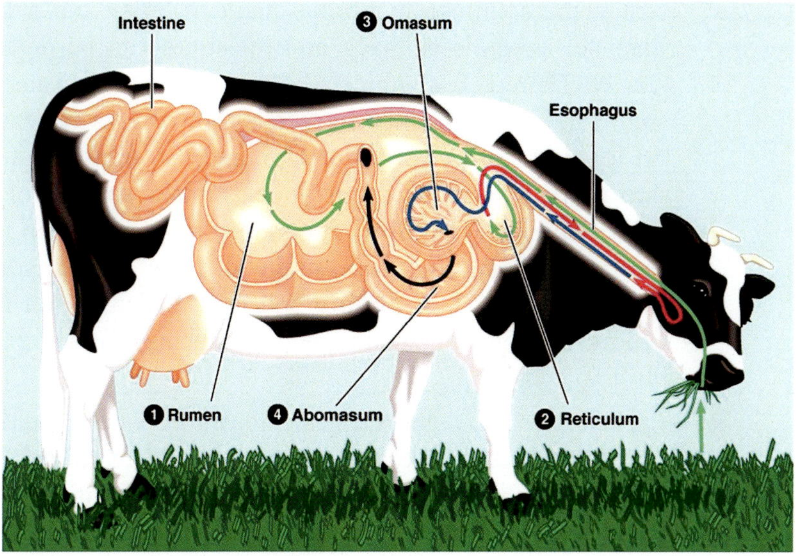

Figure 8.2. Cow's digestive system.

Source: Image by CSIRO Research.

The ultimate objective while feeding ruminants is to thus convey forage as much as possible in order to satisfy a major proportion of their nutrient's requirement. The quality and quantity of roughage that is fed to the ruminant will then govern the quantity and type of concentrate or supplement to be fed:

1. **Feeding of Young Ruminants:** In young stock, the rumen is not well developed, and it will take some months until the rumen is entirely grown and starts functioning. Until then, the young ruminant is comparable to a simple-stomached animal with context to the nutrition intake.

In young stock, critical amino acids should be offered in needed quantity in the ration. The vitamins D and A, B-complex vitamins, and minerals should be provided generally from the milk. Colostrum should be given at days 1 to 3 after birth as antibodies (gamma globulins) are transmitted from the dam to its young one.

2. **Feeding of Adult Ruminants:** It is usually seen that ruminants have a forestomach constituted of fermentation compartments, which comprises a huge quantity of microorganisms (protozoa, bacteria, fungi). These break

down the cellulose in fibrous plant material into a form that the animal's stomach and intestines can be digested easily. There is a symbiosis between microorganisms and ruminants, as the former one needs the nutrients and energy in forage for their own nutrition, and the microorganisms are ultimately shattered down as protein source for the host ruminant. Therefore, ruminants are generally in need of lesser grains and concentrate diets in comparison to monogastric such as poultry and pigs, which do not have a forestomach filled with microorganisms that serve as a protein source.

3. **Ruminant Nutrition:** In ruminant nutrition, one should be aware of the amount of energy intake needed by an animal for a specific production function, if it is preferred to attain the most efficient consumption of a feedstuff. During the process of food metabolism, energy in the diet is wrecked down from gross energy into net energy (NE) for production and maintenance.

To meet the required energy in ruminants, the feeds energy value is most imperative but one should also need to maintain the balance of other nutrients as well as such as amino acids, proteins, fats, vitamins, and proteins. It is often seen that the deficiency of any one of them will result in disruting in metabolism.

To minimize the probabilities of nutritional deficiencies, several feeding systems have been framed to assist nutritionists in selecting ration components. These systems involve:

i. Real-life application of the basic concepts of energy systems;
ii. Metabolic processes whereby energy is freed from individual nutrients;
iii. The functions performed by unstable fatty acids in ruminant nutrition.

It is worth noticing that in general, as the fibre level of ruminant rations drops, the proportion of acetic acid in the rumen matters also falls. The fibre portion of feeds is generally shattered down into acetic, butyric, and propionic acids, and about 60% of the digestible energy from fibrous carbohydrates is transformed to volatile fatty acids (VFA) within the rumen.

The conversion of carbohydrates to VFA is basically relies upon the microbes that exists in the ruminant digestive tract. The level of 8% crude protein of diets is requisite to convey the minimum ammonia (NH_3) levels needed by microorganisms for optimal rumen movement.

8.4. FORMULATION OF LEAST COST RATIONS

The aim in preparing least-cost rations, especially on enormous commercial farms, is to commence precision feeding in a way to minimize the cost and maximize the economic efficiency. In the past, there was a widely followed trend to over formulate diets when the strict necessities, particularly for critical nutrients such as phosphorus and amino acids for monogastric, were indeterminate (Figure 8.3).

Figure 8.3. Formulation of least-cost rations for cow.

Source: *Image by Progressive Dairy Solutions Limited.*

This practice is presently known to be inefficient and also result in the elimination of surplus nutrients in manure, ultimately acting as a contributor to environmental pollution.

After ascertaining the nutritional needs of a group of livestock, the next footstep would be to match these requirements with the use of a grouping of ingredients and additions to attain a balanced diet that stipulates adequate quantities of biologically available nutrients, especially for non-ruminants.

Thus, given the range of probable feedstuffs proximate composition, and the directed dietary nutrient levels anticipated, a lot of calculations are then undertaken to attain at least-cost diets.

Although, over the years, feed preparation has progressed from an ingenious balancing of little feedstuffs for a narrow number of nutrients to a linear programming system that functions with the help of computers.

8.5. SOME BIOTECHNOLOGICAL AND ALLIED APPLICATIONS EMPLOYED IN LIVESTOCK NUTRITION

It is important to note that modern biotechnology possesses the ability to offer new methods and means for attaining improved livestock productivity in ways that could lessens poverty, promote nutrition and food security, and also endorse sustainable use of natural resources (Figure 8.4).

Figure 8.4. Using applications for cow management.

Source: *Image by PGG Wrightson.*

The author noted that there could be the creation of new ingredients such as yeast protein and single-cell protein, and the ultimate objective is to produce microbial proteins as new feed sources for animal feeding.

These could also be comprised in the ration of livestock in a way to enhance the crude protein content of the ration. Furthermore, as summarized, there could be the presentation of designer ingredients that could be realistic in designing or framing GE forage plants and crops, which are genetically modified (GM) with the help of recombinant deoxyribonucleic acid (DNA) technology aiming at enhancing or introducing a necessary characteristic in the seed or plant used. It is important to note that transgenic forage crops are concerned with providing some benefits directly to consumers.

Thus, when transgenic forage crops are first consumed by ruminants, then the animal products consumed by humans from these ruminants are not themselves transgenic. This suggests that food products derived from animals fed with transgenic forage crops are much safer compared with when these modified crops are directly consumed by humans.

Furthermore, in another research, it was revealed that the addition of GM corn silage in the diets of dairy cows did not impact milk production or feed intake. The corn silage diet, which is consumed by the dairy cows was engineered with significant advancement in their nutrient (amino acids, proteins, oils, starches, fatty acids, sugars, vitamins, fibre, enzymes, minerals) contents.

The milk production or feed intake was not affected negatively, and there was a lack of transgenic DNA in the milk gathered from these experimental cows.

Therefore, plants or designer ingredients (e.g., high oil maize) with genetic modification are made to augment nutrition. There could also be designer ingredients (e.g., forage crops or low-phytate maize designed to minimize the level of antinutritive compounds that arises in livestock feed ingredients.

It was reported that feeds obtained from GM plants (25% of which are now grown in developing and less developed countries), such as silage, grain, and hay, have subsidized to an increase in milk production and livestock growth rates.

In addition, GM crops having better-quality amino acid profiles can be used to lessens nitrogen excretion in poultry and pigs. It is worth noticing that with the increase in the levels of amino acids in grains, the essential amino acid necessities of poultry and pigs can be gathered by diets that are lesser in protein proportion.

Other biotechnological applications of distinctive categories of feed additives are the use of antioxidants, crystalline amino acids, antibiotics, antifungals, and distinctive classes of antibiotic replacers. Feed additives may be supplemented to the diet to augment the efficiency of nutrients, and they also employ their results in the gut or on the gut cell walls of the animal.

They are primarily used to endorse animal growth through their effect on amassed feed palatability and quality. Besides, they are hybrid with the feed-in nontherapeutic amounts and thus shielding the animal against all kinds of injurious environmental strains. Low levels of additives in animal feed may contribute to increased production of animal protein for human consumption and thus rescue the overall cost of animal product.

The use of spiramycin, avoparcin, zinc bacitracin, tylosin phosphate and virginiamycin, as animal feed additives were prohibited in the European Union in the year 1998 and 2006. The US, at the starting of the year 2017, also imposed a sanction on the use of antimicrobials (antifungals and antibiotics) to encourage food-animal growth.

Envisaging a total ban on in-feed antibiotic use, a multitude of compounds (independently and in groups) are being tested to aid as substitutes. Probiotics can be defined as feed supplements that are given as a supplement with the diet to farm animals to enhance their intestinal microbial balance.

Therefore, contrary to the use of antibiotics as nutritional modifiers, which demolish bacteria, the addition of probiotics in feeds are intended to incite certain strains of bacteria in the gut at the cost of less wanted gut microorganisms.

It was also noted that probiotics could produce vitamins of the digestive enzymes and B complex, and the stimulation of intestinal mucosa immunity, by mounting defence against toxins formed by pathogenic microorganisms.

Thus, in ruminants, probiotics play a very critical role in controlling the diseases of the gastrointestinal tract of young animals. It was observed that in adult ruminants, yeasts can also act as a probiotic to advance the fermentation of rumen. The use of such feed additives would be useful in making animal products with better quality and more consistency.

8.6. EARLY DEVELOPMENTS IN ANIMAL FEED PRODUCTION

The first use of ground grain as the feed of animal was witnessed in the year 1813, but grinding grains for livestock served as a secondary purpose of early mills. Initially, mills were in place to grind grain for the consumption of humans, and the leftovers were naturally fed to livestock and poultry in order to prevent waste.

Oats were conventionally used for feeding animals, particularly equine, as they were inexpensive to grow and very less likely used by human as consumption in comparison to wheat. The end of the 19th century marked a termination of events that trigger the industrialization of animal feed into power. The industrial revolution had already established it's footprints in meatpacking, grain milling, and milk processing. The parallel rise in milling waste that was being discarded into water sources rapidly got negative responsiveness and was terminated by governmental rules, regulations, laws, and policies.

During this time, the importance of a balanced diet in animal production was being recognized. There was an instant demand to redirect by-products from human food production to evade waste build-up and to prove complete feeds.

When chemical analysis of these grain by-products substantiated their nutritional significance, the commercialization of such products was initiated on a large scale termed as modern feed industry. Currently, these by-products are more adequately categorized as co-products of grain and animal processing, using technologies that extract energy value and essential nutrient.

The first commercial opportunity for producing feeds on a large scale, as we would recognize them today, arose in the early 1800s when mules and horses act the core form of transportation and energy to drive agricultural equipment.

The network of livery stables that supplied rest and feed for horses were almost as ubiquitous as gasoline and service stations are today (Schoeff et al., 2005). Better quality feed for mules and horses was in great demand, and many of the major feed companies in presence today (e.g., Cargill, Purina Mills, ADM, and Ridley) started by making the production of value-added feeds for mule and horse feeds, although very little scientific feed manufacturing or formulation was involved.

Manufacturing articulated feeds grounded on the first feed standards started in the late 1800s. In 1870, the first batch of mixed feed was stated in Massachusetts, but it was only after the first nutrition books on animal were published by Henry and Armsby in the 1880s that the industry took off as a whole. The first corn gluten was produced in the year 1882, and Cargill, who was chiefly involved in flour milling, arrived in the animal feed industry just 2 years later in 1884.

In the mid-1800s, sociologists such as Mayhew (1851) were writing about the 'cat meat men' existing in large and popular cities such as London, who purveyed mostly horsemeat to households for their pets.

In 1890 meat scraps were the first one of the by-products to be renowned for their high-class nutritional value and espoused by the developing commercial feed industry. The 1890s also announced the integration of brewing by-products into animal feed and saw the start of Purina Mills in the year 1894 (Pederson, 2000).

8.7. THE U.S. FEED INDUSTRY

At the starting of the 20th century, there has been significant advancement witnessed with respect to the feed manufacturing industry. The strengthening of the food system firstly embraced the development of dedicated farming units, but the introduction of synthetic fertilizers in the 1900s aimed at controlling the levels of nitrogen in the soil and considerably advanced farm productivity (Figure 8.5).

Figure 8.5. Sustainability in animal nutrition.

Source: *Image by CM-Life Science.*

This result in better specialization and intense increases in yields of crops, resulting in a surge in livestock production. Overall, the feed industry and agriculture initiated to incorporate more advanced mechanization and technologies in the hunt for feed efficiency and uniformity. In the years starting with the 1900s, the first hammer mill was made into use, accompanied by the horizontal batch mixer in the year 1909.

The first quarter of the 20th century saw quick evolution in the mill, but the whole industry was transformed when Purina headed pelleted feed in the 1920s. This method was adopted to compress fine, variable density, unpalatable, and difficult feed resources into easy to grip pellets with higher uniformity, and was quickly embraced by other manufacturers. By the year 1930s, there were several feed millers came into existence that specialized in pellet production. In the U.S. feed mills had been conventionally situated in cities such as Minneapolis or Chicago for their vicinity to grain mills.

Companies that are already associated in the milling industry were severely intricate in the feed industry as a method to utilize grain processing by-products to use it as an alternative rather than disposing of them. The traditional feeds were produced by complementing nutrients to simple flour by-products.

Though the 1940s and 1950s marked a time when preparations became more multifaceted as critical trace minerals were recognized, synthetic vitamins were advanced, and the accumulation of antibiotics came into existence.

Concurrently, the feed industry started to expand its base beyond the original flour milling companies, with a large number of firms taking entry into the feed business; reaching a total number of more than 2,000.

Moving away from the original hubs, smaller feed mills achieved the new multifaceted diet preparations with batch mixing and were in locations closer to the farm. This change was imperative with respect to the opportunities for specialization, as was the introduction of extruded pet food later in the 1950s.

By the end of the year 1950s, specialization and advancements had sustained to favour a trend of consolidation in the feed industry. Transitioning to the 1960s and 1970s embraced the development of very large feed mills so that the norm became plants proficient enough to produce 200–500,000 tons yearly.

Relocated to large seaports, grain terminals and along railway lines and rivers, these mills were only restricted by the charges of transporting raw materials and distribution of the final product. Although, cattle feedlots as well as other large farms preferred to install their own feed mills to gain competitive strength.

Meanwhile, farm technologies such as chemical pesticides, fertilizers, hormones, and antibiotics became ordinary. The development of the feed industry headed to augmented consolidation of farms that get the benefits of economies of scale in production.

At the same time, the total number of farms plunged, while the size of the persisting farms amplified, resulting in a significant increase in the overall productivity. In fact, from 1950 to 2000 the total number of U.S. farms fall from 5 million to almost 2.2 million.

Consolidation is not the only trend in the industry that has sustained till date in the U.S. In the year 1975, enhanced automation of the feed mill was witnessed and has since progressed to curtail feed rate per unit of animal produced and maximize the production output of feed efficiently and effectively.

Since the first automated feed mill, the software and technology behind processing have persistent to improve ingredient receiving logistics, ground ingredient particle size features, mixing, and matching, extrusion processing, pellet processing, process control, post-pellet applications, feed delivery logistics and many other aspects of manufacturing (Ferket and Stark, 2011).

Over the past several years, technology has expanded in order to make the production process more automated while at the same time allowing specializing for a given feed mill's objective.

More recent advancement in equipment comprises liquid applicators in 1990, the overall mechanization of operations, data gathering in the feed mill, and near-infrared (NIR) in-line nutrient analysis.

8.8. THE FUTURE OF FEED

The future of the feed industry will likely to grow at a rapid rate with the significantly increasing global population, changing socio-economic climate, increasingly aware and active consumers and media, and the constant pursuit for greater efficiency and effectiveness. The surge in population and GDP proposes that demand for both meat protein as well as cereal grain will be expected to grow (Figure 8.6).

Figure 8.6. Future of animal nutrition.

Source: *Image by HobbyFarms.*

Therefore, what would be the future of the industry to acclimate these objectives? At this point in time, the price of feed expresses a need on the part of companies to re-assess and advance feed efficiency. Developments in raw materials may result in novel feedstocks that alternate the entire tactic to feed formulation.

This is no longer a worry only for monogastric animals. Inexpensive feed options conventionally offer to ruminants are inadequate, and the future substitute is a cautiously articulated feed targeted for greater efficiency in ruminants and monogastric. As the modern consumer has touched a new level of commercial and political involvement, there will be augmented demands on the feed industry to hold responsible for their function in the food chain.

As companies likely to be consumer-focused, so their emphasis on sustainability and their environmental footprint grows. Greenhouse gas (GHG) emission and environmental impacts of animal production will all be taken into consideration and will push the drive of feed mill and farm efficiency forward.

It is generally seen that consumers are becoming choosier in purchasing meat, eggs, and milk, inquiring what their animals eat and how they are to be kept.

They see the feed mill as essential in the food chain and shaping the human diet, which has now resided into a culture that wants to comprehend precisely what medicines, supplements, practices, and feedstuffs go into the animals they consume. Food safety will likely remain an area of major concern as today's feed may comprise hazards, such as high dioxin levels, heavy metals, mycotoxins, and PCB's. All of these toxins possess the ability to enter the food chain via animal feed. Analytical systems have been developed to detect the existence of such menaces qualitatively and quantitatively onto feed.

Over time these methods will become more urbane and provide quicker results, allowing the government and industry to react in order to ensure food safety. Food quality will likely to be the topmost priority for several companies and consumers, putting it at the frontage of feed manufacturing.

Concerns about food quality and safety will be prevalent at every stage in feed production, thus fostering more and more companies to be increasingly transparent. Government regulations, like the Food Safety Modernization Act in the U. S., continue to impose risk-based preventative animal feed safety systems to avoid the potential threat.

8.9. THE INFORMATION AGE AND THE FEED INDUSTRY

In the future, it is expected that the feed systems will likely advance further and integrate with the information technology, ultimately letting the flow of detail throughout the steps in the food chain whereby feed mill, farm, retailer, processing plant, and consumer are connected (Figure 8.7).

Figure 8.7. Making use of IT for feeding cows.

Source: *Image by Farmers Weekly.*

In this way, feed availability will be able to reflect demand and, in doing so, severely increase efficiency and minimize the overall waste.

Widespread digitalization will be helpful in monitoring and measuring the conditions and functioning of farms on a real-time basis. This includes automated weighing scales sending continuous data on body weight, feed consumption, and temperature for enhancing the efficiency of animal welfare and production, while at the same time restrictive waste.

Efficient systems will be executed in place that plays a critical role in measuring environmental gases, excretion on land and other chemical and biological parameters that will be beneficial in analyzing the impact of nutrients on the health of animals, thus offering a vision for needed changes in a diet strategy. Such measurements will also contribute to augmented on-farm biosecurity.

As new and more precise technologies are now becoming a reality to address nutritional challenges, it is well known the research on feed manufacture, nutrition, and feeding practices will continue to grow. This will be reflected in the overall strategies, structure, and activities of the global feed industry.

It is conceivable to witness these changes taking place today, as commercial and research applications focus more on modelling the dynamics of nutrient digestion, utilization, and balance, transitioning from feeding practices based on the simple nutrient composition.

As scientists are now coming up with more advanced technologies for nutrient utilization and controlling digestion, conventional ways of articulating feeds on the basis of compositional analyzes will play an insignificant role in defining feed formulation.

These fine control mechanisms will be paired with our potential to amass information and data about all aspects of the food chain and will outline a new horizontal of proactive precision feeding. It will be imperative to abandon some of the time-honoured ways of defining nutritional value.

This will provide the opportunity to make use of new feed ingredients, long-term nutritional preparing and other innovative supplement strategies. The outcomes will comprise better maintenance of animals, better production efficiencies, and a more sustainable production process.

Nutrigenomics is already being employed in the growth and understanding of certain nutrients and is projected to offer standards for understanding both environmental as well as nutritional factors deterring production welfare and efficiency. Such information will play a very useful role in a new level of precision in livestock production and feed preparation.

Nutrigenomics is very critical in evaluating the influences of distinctive nutrients on gene expression and act as a foundation for more rational genetic selection. Consequently, feeds will be closer to attaining genetic potential by targeting genes that play an active role in animal growth rate, meat quality, and disease prevention.

These studies are useful in identifying anti-nutritional activity and outline feeding strategies that take benefit of the natural conditioning processes concerned with perinatal and prenatal feed management practices.

The feed mill, itself, will become 'smarter,' for instance, with the use of NIR technology. This allows making evaluation of entering raw materials on a real-time basis, allowing reformulation of the diets on a minute-by-minute basis to certify the reliability of each batch produced.

In addition, rapid, *in vitro* digestion modelling systems will offer new ways of outlining the actual nutritional significance of raw materials as well as the final feed products leaving the feed mill. These advanced systems would be useful in expressing nutrient interactions and provide more precise narratives of nutritional values.

Real-time decisions and Precision modelling, resulting from better sums of information, will ensure that the feed industry will be able to address the issues related to sustainability as well as minimize waste and the environmental impact from animal production. At the end, such technologies have the potential to offer a more objective feedstuff value for the interested buyer (Gill, 2003).

Today, a typical western feed mill having a production capacity of 100,000 tons of feed might have employees of not more than three people. In countries such as China, the same feed mill would require almost 45 employees.

This scenario is projected to sustain in the future, whereby feed mills would become fully automated, and there will be a requirement of only a single person to supervise the entire automated process in a mill efficiently and effectively.

As mill quality control systems are expected to further via automation for each of the distinctive steps of production, traceability technology will facilitate to ensure that larger volumes being tracked and transported, while at the same time reducing the labour size.

Not only will it incorporate a greater level of efficiency, but the increase in data collection and precision can then increase exponentially. In the future, there is potential to make use of artificial intelligence in dairy farms to interpret and analyze data and make instant adjustments on the basis of system inputs.

Advancements are expected to arise in the pelleting process so that what is now pondered an art might become more of a science. By taking into consideration the majority of the parameters involved in cooling and pelleting, there is an opportunity to improve the quality of feed, reduce the overall costs, and assure food safety.

8.9.1. Key Ingredients

The degree to which the above-described solutions can be adopted by the industry is based upon the several challenges faced by them. However, the instant approach of the feed industry is to advance the overall performance of the animal, minimize the associated costs, and maximize feed production efficiencies. Above all, FCR and digestibility need consideration. Novel ingredients would likely be studied for their capacity to effectively improve FCR. Once identified and advanced, such ingredients become essential insertions in diet preparations, particularly in situations when the competence of each point of feed conversion becomes highly imperative.

One probable ingredient that could become ordinary to feed formulation is algae: a source of protein, pigments, oils, starch, and vitamins. Developing technology made algae become a chief, nutritionally rich biomass and it is already being recognized as a prime source of omega-3 fatty acids. Initial trials exploiting algae produced results showing progress in both the nutritional values of meat and reproductive performance.

Solid-state fermentation (SSF) is a conventional technology that, with potential applications and techniques, can be used for growing enzymes straightaway on feedstuffs. SSF provides the opportunity to create a custom enzyme cocktail that can be used for enhancing digestion of feedstuffs, which improves the overall health of animal, profitability, and performance.

The process comprises developing a specific strain of Aspergillus fungi rightly on feed, which gives a substrate-specific array of enzymes that assists in the digestion of fibre and other essential nutrients to an extent that has never been attained with conventional, batch fermentation methods. Data from trials in pigs, poultry, and aquaculture have shown better digestion and noteworthy savings as a result of embracing such products in the feed.

8.10. CONCLUSION

At the end, it is concluded that animal nutrition is one of the most factors that should be taken care of. With the increase in the demand for poultry products, providing adequate feed to the animal is the most significant task.

While offering feeds to the animals, several things need to take care of. First and foremost is to make sure that it should include antibiotics and the bacteria in the feeds should not be antibiotic-resistant. In addition, animals should give feeds in adequate quantity to provide them with proper growth and development.

In the future, there is a need to align technology in the feed industry as it will help in providing nutrition to animals in adequate quantity and quality. Technology will help to develop balance and cooperation between various stakeholders in the animal feed industry. It helps in connecting feed mill, farm, retailer, processing plant, and consumers.

Also, as it is well known that every livestock has different needs with respect to nutrient level. Therefore, it is the duty of the people who deal with animals to carefully assess the nutritional needs of animals as the need of one animal may be different from another. Therefore, developing a proper diet chart for each animal will work best in this case.

REFERENCES

1. Animalscience.tamu.edu. (2018).*Animal Nutrition | Animal Science.* [Online] Available at: https://animalscience.tamu.edu/academics/nutrition/ (accessed on 15 March 2021).
2. Arai, T., (2014). The development of animal nutrition and metabolism and the challenges of our time.*Frontiers in Veterinary Science*, 1.
3. Breeding & Genetic Selection, (2021).*Breeding and Genetic Selection.* [Online] Available at: https://www.uaex.edu/farm-ranch/animals-forages/beef-cattle/breeding-genetic-selection.aspx (accessed on 15 March 2021).
4. Cherian, G., (2019).*A Guide to the Principles of Animal Nutrition.* [Online] Open.oregonstate.education. Available at: https://open.oregonstate.education/animalnutrition/ (accessed on 15 March 2021).
5. Cherian, G., (2021).*A Guide to the Principles of Animal Nutrition.* [Online] Open.oregonstate.education. Available at: https://open.oregonstate.education/animalnutrition/ (accessed on 15 March 2021).
6. Der, P. A., Abdollahi, M., Cheng, H., Colovic, R., Den, H. L., Miladinovic, D., Page, G., et al., (2020). Future directions of animal feed technology research to meet the challenges of a changing world. *Animal Feed Science and Technology,270*, 114692.
7. Ferket, P., and Stark, C. (2011). Recent Developments in FeedTechnology.North Carolina State University Dept of Poultry Science report.
8. Gill, C. (2003). Grain grading going digital.Feed International23:6–9.
9. Godfray, H. J., (2021).*Meat: The Future Series Alternative Proteins.* [eBook] Available at: http://www3.weforum.org/docs/WEF_White_Paper_Alternative_Proteins.pdf (accessed on 15 March 2021).
10. Hammack, S., (2021).*Texas Adapted Genetic Strategies for Beef Cattle V: Type and Breed Characteristics and Uses*. [eBook] Available at: https://animalscience.tamu.edu/wp-content/uploads/sites/14/2012/04/geneticsE190.pdf (accessed on 15 March 2021).
11. Makkar, H. P. S., & Ankers, P., (2021).*Towards a Concept of Sustainable Animal Diets*. [eBook] Available at: http://www.fao.org/3/a-i4146e.pdf (accessed on 15 March 2021).
12. Mavromichalis, I., (2021).*The Future of the High-Margin Animal Nutrition Industry*. [Online] Feed Strategy. Available at: https://www.feedstrategy.com/blog/the-future-of-the-high-margin-animal-nutrition-industry/ (accessed on 15 March 2021).

13. Mayhew, H. (1851) In:Labourand labouring classes. London Spring Books, UK.
14. MLA Corporate, (2021).*Breeding Objectives and Selection.* Meat & Livestock Australia. [Online] Available at: https://www.mla.com.au/research-and-development/Genetics-and-breeding/breeding-objectives-and-selection/ (accessed on 15 March 2021).
15. Munoz, J., (2021).*The Animal Nutrition of the Future-Animal Nutrition.* [Online] Animal Nutrition. Available at: https://theanimalnutrition.com/the-animal-nutrition-of-the-future/ (accessed on 15 March 2021).
16. Nayek, S., (2019).*Future of The Global Feed Industry.* Benison Media. [Online] Benisonmedia.com. Available at: http://benisonmedia.com/future-of-the-global-feed-industry/ (accessed on 15 March 2021).
17. Orlindo, T. L., Alves, F. M., Pierre, M. J., Poppi, D. P., Carstens, G., Angerer, J., & Gene, F. D., (2021).*A Glimpse of the Future in Animal Nutrition Science. 2. Current and Future Solutions.* [Online] www.researchgate.net/. Available at: https://www.researchgate.net/publication/318557913_A_glimpse_of_the_future_in_animal_nutrition_science_2_Current_and_future_solutions (accessed on 15 March 2021).
18. Marquardt, R.R., & Brufau, J., (2021). *Future of Feed Enzymes: Orientation and Perspectives*. [eBook] Available at: References Animalscience.tamu.edu. (2018). *Animal Nutrition | Animal Science.* [Online] Available at: https://animalscience.tamu.edu/academics/nutrition/ (accessed on 15 March 2021).
19. Pederson, J. (2000). Purina Mills, Inc. History. InInternational directory ofcompany histories. Vol. 32. Detroit, Mich.: St. James Press.
20. Schoeff, R.W., Fairchild, F.J., Bursiek, B., & Castaldo, D.(2005).History of the formula feed industry. In: E.K. Schofield (Ed.) FeedManufacturing Technology V. American Feed Industry Association,Arlington, VA, USA.
21. Smith, S., (2021).*1/3: Why pay attention to Animal Feed.* [Online] Forum for the future. Available at: https://www.forumforthefuture.org/blog/why-pay-attention-to-animal-feed (accessed on 15 March 2021).
22. Tedeschi, L., Fonseca, M., Muir, J., Poppi, D., Carstens, G., Angerer, J., & Fox, D., (2017). A glimpse of the future in animal nutrition science. 2. Current and future solutions.*Revista Brasileira De Zootecnia,46*(5), 452–469.

23. Thornton, P., (2010). Livestock production: recent trends, future prospects.*Philosophical Transactions of the Royal Society B: Biological Sciences, 365*(1554), 2853–2867.

24. Tona, G., (2018). Current and future improvements in livestock nutrition and feed resources.*Animal Husbandry and Nutrition*.

25. Tona, G., (2021).*Current and Future Improvements in Livestock Nutrition and Feed Resources*. [Online] cdn.intechopen.com. Available at: https://cdn.intechopen.com/pdfs/58928.pdf (accessed on 15 March 2021).

26. Wanapat, M., Cherdthong, A., Phesatcha, K., & Kang, S., (2015).*Dietary Sources and Their Effects on animal Production and Environmental Sustainability*. [Online] ScienceDirect. Available at: https://www.sciencedirect.com/science/article/pii/S2405654515300160 (accessed on 15 March 2021).

INDEX

E

F

G

H

I

L

V

W

Kolibri 360 GmbH
Ferdinand-Jühlke-Straße 7
99095 Erfurt, Deutschland
produktsicherheit@kolibri360.de